Student Study Guide

Introduction to

Chemical Principles

Student Study Guide

Introduction to

Chemical Principles

H. Stephen Stoker

Weber State College, Ogden, Utah

prepared by
Owen C. Gayley
San Antonio College
Ruth Sherman
Los Angeles City College

Macmillan Publishing Co., Inc.
NEW YORK

Collier Macmillan Publishers
LONDON

Macmillan Publishing Co., Inc.
866 Third Avenue, New York, New York 10022

Collier Macmillan Canada, Inc.

ISBN: 0-02-417610-9

Printing: 2 3 4 5 6 7 8   Year: 4 5 6 7 8 9 0 1

ISBN   0-02-417610-9

PREFACE

To the Student:

This study guide is designed to help you evaluate whether you "really understand" the main ideas and principles found in H. Stephen Stoker's textbook PRINCIPLES OF CHEMISTRY. It is not intended that the study guide serve as a substitute for the textbook or your instructor. It is hoped, however, that it will "compliment" both the textbook and the instructor.

The organization of the study guide is the same as that of the textbook relative to chapter and section numbering.

Each study guide chapter has five sections:

     (1) Review of Chapter Objectives
     (2) Problem Set
     (3) Multiple Choice Exercises
     (4) Answers to Problem Set
     (5) Answers to Multiple Choice Exercises

A study guide chapter's REVIEW OF CHAPTER OBJECTIVES section summarizes the major concepts and problem-solving techniques found in the corresponding textbook chapter. Each review section is based on the chapter objectives, which are found in the textbook immediately preceding each chapter's problem set.

Each study guide chapter's PROBLEM SET section focuses on the main types of problem-solving techniques you should have learned from the corresponding textbook chapter. These problem sets should serve as "self-checks" on your newly-acquired problem-solving capabilities. Detailed solutions to each problem in a study guide problem set are given in the ANSWERS TO PROBLEM SET section found at the end of the study guide chapter. Students often think that they understand text material to a greater depth than they really do. Working the study guide problem sets will enable you to pinpoint any "deficiencies" in your understanding of a given chapter.

The MULTIPLE CHOICE EXERCISES provide you with a set of sample examination questions to help you determine, again, what you do and do not know. Studying these questions should

provide you with a useful "review system." Such review
should be particularly beneficial when you are studying for
an actual examination. Answers to a study guide chapter's
multiple choice exercises are found in the ANSWERS TO
MULTIPLE CHOICE EXERCISES section at the end of each study
guide chapter.

CONTENTS

CHAPTER 1
The Science of Chemistry

REVIEW OF CHAPTER OBJECTIVES

1. <u>Understand the relationship between science as a whole
   and various scientific disciplines</u> (section 1.1).

   Science as a whole is the study of man and all of his
   surroundings.

   Scientific disciplines are artificial subdivisions of
   science created by man to aid in his study and
   understanding of himself and his surroundings. It is
   important that you realize that there is no clear cut
   dividing line between one scientific discipline and
   another, but that they are all interrelated as seen in
   Figure 1-1 in your textbook.

2. <u>Give examples of how chemistry directly affects your life</u>
   (section 1.2).

   Chemistry touches all parts of our lives every day. Some
   examples are synthetic fibers used in the clothes you
   wear, such as nylon, fertilizer used to produce the food
   you eat, gasoline used to power your car and drugs used
   to treat your illnesses. Many more examples are listed
   on pages 2 and 3 of your textbook.

3. <u>Outline the major steps in applying the scientific method
   to a problem</u> (section 1.3).

   The scientific method is a set of specific procedures for
   acquiring knowledge and explaining phenomena.

   The major steps are:

   1. Collecting data through observation and
      experimentation.

   2. Analyzing and organizing the data in terms of

generalizations that summarize the experimental observations.

3. Suggesting probable explanations for the generalizations.

4. Experimenting further to prove or disprove the proposed explanation.

4. <u>Explain the difference between a fact, a law, a hypothesis and a theory</u> (section 1.4).

A fact is a valid observation about some natural phenomenon.

A law is a generalization that summarizes in a concise way facts concerning natural phenomena.

A hypothesis is a tentative model or picture that offers an explanation for a law.

A theory is a hypothesis that has been tested and validated over long periods of time.

Figure 1-2 in your textbook (page 8) summarizes the relationship of a fact, a law, a hypothesis and a theory.

PROBLEM SET

1. Indicate whether each of the following statements is true or false:

   a) a hypothesis that survives extensive testing is called a theory
   b) it is much easier to disprove a hypothesis than prove it
   c) obtaining positive results from an experiment set up to verify a hypothesis "proves" the hypothesis
   d) many scientific theories undergo modification at one time or another
   e) the minimum number of supportive experiments required to give a hypothesis theory-status is ten

2. Classify each of the following statements as a fact, law or hypothesis:

   a) maximum gas mileage is obtained when automobiles are

driven 45 mph

   b) radar indicated the car was being driven at 54 mph

   c) blue is the most common color for American-made
automobiles because more drivers are probably male
than female

MULTIPLE CHOICE EXERCISES

1. Indicate the missing words in the following statement:
"A _____ is a tentative model or picture that offers
an explanation for a _____."

   a) hypothesis and theory
   b) theory and law
   c) hypothesis and law
   d) theory and hypothesis

2. A summary statement of a large number of facts is
referred to as a:

   a) law
   b) theory
   c) hypothesis
   d) explanation

3. Indicate the missing word(s) in the following statement:
"A _____ is a hypothesis that has been tested and
validated over long periods of time."

   a) theory
   b) law
   c) fact
   d) scientific discipline

4. Which one of the following pairings of terms is correct:

   a) hypothesis -- generalization
   b) theory -- tentative explanation
   c) fact -- observation
   d) law -- validated hypothesis

5. Which of the following is not a step in the scientific
method:

   a) collecting information through observation
   b) proposing a hypothesis
   c) discarding data not consistent with a hypothesis
   d) using generalizations to summarize experimental data

ANSWERS TO PROBLEMS SET

1. a) true
    b) true
    c) false
    d) true
    e) false

2. a) law
    b) fact
    c) hypothesis

ANSWERS TO MULTIPLE CHOICE EXERCISES

1. c

2. a

3. a

4. c

5. c

CHAPTER 2
Numbers From Measurements

REVIEW OF CHAPTER OBJECTIVES

1. Illustrate the difference between the precision and accuracy of a measurement (section 2.1).

    Precision is how close a series of measurements made on the same object come to each other, while accuracy is how close a single measurement (or the average value of a group of measurements) comes to the correct value.

    An example of good precision would be if three students, asked to obtain the weight of an object, obtained the following results:

Student A	12.5 pounds
Student B	12.4 pounds
Student C	12.6 pounds

    This illustrates good precision since the greatest deviation of any one weight from another is two tenths of one pound.

    An illustration involving accuracy would be if two students were asked to weigh an object (with an accurately known weight of 4.5 pounds) and obtained the following results:

Student A	4.6 pounds
Student B	3.5 pounds

    The weight obtained by student A shows good accuracy since it is only one tenth of a pound away from the correct value. The weight obtained by student B shows poor accuracy since it is a full pound away from the correct value.

    It is important to realize that it is possible to have good accuracy with poor precision, good precision with poor accuracy, poor precision and poor accuracy, and good precision and good accuracy. Examples of this can be seen in Figure 2-1 on page 8 of your textbook.

2. <u>Determine the number of significant figures in a number resulting from a measurement</u> (section 2.2).

All numbers resulting from a measurement contain a degree of uncertainty. The degree of uncertainty is determined by the precision of the measuring device used and the skill of the person using the device.

It is by the use of significant figures that we indicate this degree of uncertainty in a measurement to other people who do not know the precision of the device used or the skill of the person using the device.

Significant figures are an important concept which many students find difficult. The key to successful application of significant figures is a thorough understanding of the rules given on pages 12 and 13 of your textbook.

Zeros cause students the most difficulty since they may or may not be significant, while the numbers 1 through 9 are always significant.

Use of the significant figure rules is illustrated in the following examples:

1. 10.50 -- 4 significant figures
    The one and five are found to be significant by applying rule one. The zero between the one and five is significant due to rule two. The zero after the five is significant because of rule four.

2. 120 -- 2 significant figures
    The one and two are significant due to rule one. The zero is not significant as a result of rule five.

3. 0.040 -- 2 significant figures
    The zero after the decimal point and before the four is not significant by rule three. The four is significant by rule one. The zero to the right of the four is significant by rule four.

You should commit the five rules on pages 12 and 13 to memory and be able to apply them correctly to all numbers which contain a degree of uncertainty.

3. <u>Identify numbers which have an unlimited number of significant figures and explain why they do</u> (section 2.2).

Numbers which result from the counting of objects or are defined to equal a definite quantity contain an unlimited number of significant figures. You can count the number of objects with absolute certainty. This means that if you count 100 chairs in a classroom there are exactly 100 and not 99 or 101. Notice that even though there are zeros in the number there is no uncertainty, and the number would contain an unlimited number of significant figures.

An example of a defined number would be twelve inches in a foot. There are always exactly twelve inches in a foot, never more and never less. Therefore, the number 12 contains an unlimited number of significant figures.

4. <u>Distinguish between significant and nonsignificant zeros in numbers containing zeros</u> (section 2.2).

The successful performance of this objective depends on the application of the rules for significant figures found on pages 12 and 13 in your textbook.

Rule 2 states that zeros between two significant figures are always significant as in 10.01. Both of these zeros are significant since they are between the two ones which are both significant numbers by rule 1.

Rule 3 states that in a number less than one, if there are no numbers found to the left of the decimal point, the zeros used to fix the decimal point are never significant, as in 0.0001. The zero found to the left of the decimal point is not significant or necessary.

Scientists usually put this zero down to prevent someone from misreading a number. For example, if the number one-tenth were written as .1, someone could overlook the decimal point and read it as 1 (one). However, if it were written as 0.1, the preceding zero calls attention to the decimal, and there is less chance of overlooking the decimal point and misreading the number. The three zeros to the right of the decimal and before the one in the number 0.0001 are not significant as a result of rule 3.

Rule 4 states that when a number has a decimal point, zeros to the right of the last nonzero are significant. The zero to the right of the decimal in the number 120.0 is significant by rule 4. If the zero to the right is significant, then the zero between the decimal and the 2 is significant by rule 2.

Notice that in rule 4 there is no mention of the position of the decimal point or the zero. All it states is that if there is a decimal point, then all zeros to the right of it and after a number are significant. These zeros are significant since they indicate the precision of the measuring device to that decimal place and must be written to indicate that the value determined was zero and not some other number. Many students tend to think this type of zero is unimportant and mistakenly do not write them down.

Rule 5 states when a number without a decimal point being explicitly shown ends in one or more zeros, the zeros that end the number may or may not be significant. These zeros cause the most difficulty since they may or may not be significant. In this book, to remove the uncertainty, such zeros will never be considered significant. For example, in the number 4300, the zeros are not significant.

5. <u>Round off measured values to a specified number of significant figures</u> (section 2.3).

The method of discarding unwanted or unnecessary numbers to obtain the correct number of significant numbers is called rounding off. It is necessary that you learn and be able to correctly apply the rules for rounding off found on pages 14 and 15 of your textbook. Digital read-outs on electronic calculators most often display more digits than are needed; hence, such numbers must be rounded off. The following examples illustrate the use of the "rounding off rules."

1. Round off the number 486.52 to three significant figures.

   Rule 2 applies here. The last digit retained (the 6) is increased in value by one unit and 486.52 becomes 487. since the first digit dropped is a 5 followed by a nonzero digit.

2. Round off the number 46329 to three significant figures.

   Rule 1 applies here since the first digit to be dropped (the 2) is less than five. That two and all following digits are simply dropped and 46329 becomes 56300. Remember significant figure rule 5 which states that zeros ending a number without a decimal point explicitly shown are never significant; hence,

46300 has only 3 significant figures. Notice also that the nine has no effect on the final result. It is only the digit which comes directly after the digit to be retained that is important when rounding, except if it is a five. Then either rule two or three applies.

3. Round off the number 8.645 to three significant figures.

Rule 3 applies; the first digit to be dropped is a 5 not followed by any other digit. The last digit to be retained (the 4) is even, so using the odd-even rule, the 5 is simply dropped and 8.645 becomes 8.64.

6. Add, subtract, multiply, and divide numbers, giving your answer to the correct number of significant figures (section 2.3).

The significant figures in a number indicate the precision of the measurement. It is important to realize that the precision of a measurement cannot be increased or decreased by a mathematical operation. Another way of stating the preceding concept is to say that the number of significant figures in a number cannot be increased by performing a mathematical operation.

There are two separate rules which are used to determine the number of significant figures allowable in a calculated result. One rule covers multiplication and division, and the other covers addition and subtraction.

For multiplication and division an answer cannot have any more significant figures than the least number present in any of the numbers which were multiplied or divided. The following examples illustrate the use of this rule.

1. 2.458 x 0.210 = 0.51618 (calculator answer)
   = 0.516 (correct answer)

The number 2.458 contains 4 significant figures and the number 0.210 contains 3 significant figures. According to the rule, the answer cannot contain more than 3 significant figures. The calculator answer of 0.51618 must, therefore, be rounded using the rules for rounding to obtain 0.516 for the correct answer.

2. 2.0 x 5.16 x 4.189 x 89700 = 3877774.1 (calculator answer)
   = 3900000 (correct answer)

You should see from this example that no matter how many numbers are multiplied or how many significant figures any of the numbers contain, the answer can contain no more than the least number. In this example, the least number of significant figures is 2 in the 2.0, so the correct answer is 3900000. It is may be difficult to understand this, but if you recall the concept of precision and understand the meaning of significant figures, this is the only logical answer. Even though 3877774.1 seems to be much more exact it is an incorrect answer.

For addition and subtraction, the answer cannot have digits beyond the last digit position common to all numbers being added or subtracted.

Example:

$$14.46 + 8.3 = 22.76 \text{ (calculator answer)}$$
$$= 22.8 \text{ (correct answer)}$$

The digit position common to both numbers is the tenths place. Therefore, the calculator answer of 22.76 must be rounded to 22.8 since it cannot contain any digits past the tenths place. Notice that 8.3 only contains 2 digits, but the correct answer 22.8 contains 3. Students who apply the rule for multiplication and division to this problem would say the answer should be 23. This is incorrect according to the rule for addition and subtraction. It is important to remember there are two different rules and to know when to apply them.

7. **Convert numbers from decimal notation to scientific notation and vice versa** (section 2.4).

Scientific notation is a system for expressing large or small decimal numbers as numbers between 1 and 10 multiplied by 10 raised to a power. The correct number of significant figures is carried in the number between 1 and 10.

If a decimal number is larger than one, the decimal point is moved to the left until the number is between 1 and 10. The power of 10 used with the number is equal to the number of places that the decimal point was moved to the left.

Example:

1. $6874 = 6.874 \times 10^3$

The number is larger than one so the decimal point is moved to the left three places to obtain a power of 3. The number contains 4 significant figures and all 4 are retained.

2. $6,830,000 = 6.83 \times 10^6$

The number is larger than one so the decimal point is moved to the left 6 places to obtain a power of 6. The four zeros are not significant and are not carried when the number is written in scientific notation. This is one method of discarding zeros which are not significant.

If the number is smaller than one, the decimal point is moved to the right until the number is between 1 and 10. The power of 10, the number of places the decimal point was moved to the right, is a negative number. (When the decimal point movement is to the left the power of ten is always positive; when movement is to the right the power of ten is always negative.)

Example:

1. $0.043 = 4.3 \times 10^{-2}$

The number is smaller than one so the decimal point is moved to the right 2 places and the power is −2.

2. $0.0010 = 1.0 \times 10^{-3}$

The number is smaller than one so the decimal point is moved to the right 3 places and the power is −3. Notice that zero to the right of the one is written since it is significant and must be carried in the number between 1 and 10. This type of zero is often incorrectly left off by students.

To convert a number in scientific notation into a decimal number, the preceding processes are reversed.

If the exponent is positive, the decimal point is moved that number of places to the right. Zeros may have to be added to the number as the decimal point is moved.

Example:

1. $4.647 \times 10^2 = 464.7$

The exponent is a positive 2, so the decimal is moved 2 places to the right.

2. $4.6 \times 10^5 = 460,000$

The exponent is a positive 5, so the decimal point is moved 5 places to the right. Four zeros have to be added so the number is of the correct magnitude.

If the exponent is negative, the decimal point is moved to the left.

Example:

$2.64 \times 10^{-3} = 0.00264$

The exponent is a negative 3 so the exponent is moved 3 places to the left.

8. <u>Carry out mathematical operations with all numbers expressed in scientific notation</u> (section 2.5).

In multiplication, the coefficients of scientific notation numbers are multiplied and the exponents are added algebraically.

Example:

1. $(3.3 \times 10^3) \times (8.1 \times 10^2) = 26.73 \times 10^5$
$$\text{(calculator answer)}$$
$$= 27 \times 10^5$$
$$= 2.7 \times 10^6$$
$$\text{(correct answer)}$$

The coefficients are 3.3 and 8.1. When they are multiplied together, the result is 26.73. Since 3.3 and 8.1 both contain 2 significant figures, the product is rounded off to 27.

The exponents are 2 and 3. The sum of 2 and 3 is 5. Therefore, 10 is raised to the 5th power in the answer to give 27×10^5.

The answer 27×10^5 must be justified to obtain the correct answer of 2.7×10^6.

(If you have a calculator that will handle numbers in scientific notation the above calculation would have been a 2 step process rather than 3 step. The calculator answer would have been 2.673×10^6 since the calculator automatically justifies scientific notation numbers. Rounding this answer to 2 significant figures gives 2.7×10^6.)

2. $(6.12 \times 10^{-2}) \times (3 \times 10^{-4}) = 18.36 \times 10^{-6}$
$$\text{(calculator answer)}$$
$$= 20 \times 10^{-6}$$
$$= 2 \times 10^{-5} \text{ (correct answer)}$$

The coefficients 6.12 and 3 are multiplied to obtain 18.36. There is only one significant figure in 3 so 18.36 is rounded to 20.

The exponents are -2 and -4. The algebraic sum of -2 and -4 is -6. Therefore, 10 is raised to the negative 6th power in the answer to give 20×10^{-6}.

The 20×10^{-6} is then justified to obtain the correct answer of 2×10^{-5}.

3. $(4.056 \times 10^{-4}) \times (1.34 \times 10^2) = 5.43504 \times 10^{-2}$
$$\text{(calculator answer)}$$
$$= 5.44 \times 10^{-2}$$
$$\text{(correct answer)}$$

The coefficients 4.056 and 1.34 are multiplied to obtain 5.43504. Since 1.34 contains 3 significant figures, the 5.43504 is rounded to 3 significant figures to give 5.44.

The exponents are -4 and 2. The algebraic sum of -4 and 2 is -2. Therefore, 10 is raised to the negative 2nd power in the answer to give 5.44×10^{-2}.

In division, the coefficients of the exponential number are divided (the numerator must be divided by the denominator) and the exponents are algebraically subtracted (numerator exponent minus the denominator exponent).

Example:

$$\frac{5.34 \times 10^2}{7.6 \times 10^{-3}} = 0.70263158 \times 10^5 \text{ (calculator answer)}$$
$$= 0.70 \times 10^5$$
$$= 7.0 \times 10^4 \text{ (correct answer)}$$

The coefficients 5.34 and 7.6 are divided to obtain 0.70263158. There are 2 significant figures in 7.6

so 0.70263158 is rounded to 0.70.

The exponents are algebraically subtracted to obtain 5. Algebraic subtraction involves changing the sign of the number to be subtracted (the exponent in the denominator) and then following the rules for addition. Therefore, 10 is raised to the 5th power to give 0.70×10^5. The 0.70×10^5 is then justified to obtain the correct answer of 7.0×10^4.

In addition and subtraction, all numbers must be expressed in terms of the same power of 10. When all the numbers have been expressed with the same exponent, the numbers can then be added or subtracted and the exponent maintained at its now common value.

Example:

1. $(4.35 \times 10^3) + (3.34 \times 10^4) = (0.435 \times 10^4)$
    $+ (3.34 \times 10^4) = 3.775 \times 10^4$ (calculator answer)
    $= 3.78 \times 10^4$ (correct answer)

    Change 4.35×10^3 to the same power of 10 as 3.34×10^4; 4.35×10^3 becomes 0.435×10^4.

    Add the coefficients of the numbers that are multiplied by the same power of 10 and keep the power of 10 to obtain 3.775×10^4, the calculator answer.

    The calculator answer is rounded to the tenths place (the last common digit) to obtain the correct answer of 3.78×10^4.

2. $(7.2 \times 10^3) - (4.9 \times 10^2) = (7.2 \times 10^3) - (0.49 \times 10^3)$
    $= 6.71 \times 10^3$ (calculator answer)
    $= 6.7 \times 10^3$ (correct answer)

    Change 4.9×10^2 to the same power of 10 as 7.2×10^3; 4.9×10^2 becomes 0.49×10^3.

    Subtract the coefficients of the numbers that are multiplied by the same power of 10 and keep the power of 10 to obtain 6.71×10^3, the calculator answer.

    The last digit place common to both numbers if the tenths place so 6.71×10^3 is rounded to 6.7×10^3.

    The result is already justified so the correct answer is 6.7×10^3.

In calculations involving both multiplication and division it is usually easiest to simplify the numerator and denominator into single numbers before performing the division step.

Example:

$$\frac{(6.34 \times 10^3) \times (5.42 \times 10^4)}{(9.20 \times 10^2) \times (3.64 \times 10^6)} = \frac{34.4 \times 10^7}{33.5 \times 10^8}$$

$$= 1.0268657 \times 10^{-1} \text{ (calculator answer)}$$

$$= 1.0 \times 10^{-1} \text{ (correct answer)}$$

The 6.34×10^3 and 5.42×10^4 in the numerator are multiplied and the result rounded to 3 significant figures to obtain 34.4×10^7.

The 9.20×10^2 and 3.64×10^6 in the denominator are multiplied and rounded to 3 significant figures to obtain 33.5×10^8.

The 33.5×10^8 of the denominator is then divided into the 34.4×10^7 of the numerator to obtain the final answer.

PROBLEM SET

1. The weight of an object is known to be 24.6445 pounds. Three students, A, B and C, are asked to determine the weight of the same object using a less precise scale. Each student weighs the object six times with the following results:

    A. 24.4, 24.5, 24.6, 24.7, 24.7, 24.7 (average = 24.6)
    B. 25.6, 23.8, 24.1, 24.6, 25.0, 25.2 (average = 24.7)
    C. 24.8, 24.9, 25.2, 24.8, 25.0, 25.1 (average = 25.0)

    Rate each student's performance in terms of accuracy (good or poor) and precision (good or poor).

2. Determine the number of significant figures in each of the following measured values:

    a) 0.020   e) 1.000
    b) 3.01   f) 2.050
    c) 99.8   g) 153.70
    d) 4200   h) 0.0006

3. How many significant digits are there in each underlined number? If there is an infinite number, explain why.

   a) 1 gross = 144   c) 1 quart = 32 ounces
   b) 6 eggs   d) body temperature = 98.6° F

4. How many nonsignificant zeros are present in each of the following numbers?

   a) 11.0   d) 5280
   b) 0.045   e) 12,600,000
   c) 206

5. Round off each of the following numbers to 3 significant figures:

   a) 6463   e) 984.52
   b) 1.475   f) 15.97
   c) 4.896   g) 0.03010
   d) 64.25   h) 0.103566

6. Carry out the following multiplications, expressing your answers to the correct number of significant figures. Assume that all numbers arise from measurements.

   a) 5.87 x 2.0   e) 5 x 3
   b) 2.004 x 3.0   f) 7 x 4.68795
   c) 4.8777 x 3.6555 g) 5200 x 8.4765
   d) 2.856874 x 5.8 h) 0.024 x 1.005

7. Carry out the following divisions, expressing your answers to the correct number of significant figures. Assume that all numbers arise from measurements.

   a) $\dfrac{3860}{2.2222}$ e) $\dfrac{3800.0}{4.0000}$

   b) $\dfrac{2000}{1.2}$ f) $\dfrac{2}{30}$

   c) $\dfrac{9}{2.0}$ g) $\dfrac{2.01265}{3}$

   d) $\dfrac{4.00}{2.02}$ h) $\dfrac{1,000}{6.245}$

8. Carry out the following mathematical operations, expressing your answers to the correct number of significant figures. Assume that all numbers arise from measurements.

a) $\dfrac{3.0 \times 9.0}{2.2}$

c) $\dfrac{3.586}{2.02 \times 3.02}$

b) $\dfrac{2.222 \times 3.333}{6.6}$

d) $\dfrac{5.560 \times 2.002}{0.02123 \times 0.0023}$

9. Perform the following addition-subtraction computations. Report your results to the proper number of significant figures. Assume that all numbers are measured numbers.

   a) 3 + 7 + 11
   b) 3.0 + 7.00 + 32.00
   c) 4.28 − 4.2
   d) 4.344 − 3.6809

   e) 888.0 + 0.6 + 0.26
   f) 2300.0 + 42 + 2.026
   g) 0.1206 − 0.00306
   h) 60.0 + 0.00000440

10. Express the following numbers in scientific notation:

   a) 345
   b) 0.00428
   c) 40.4003
   d) 2,000

   e) 0.340
   f) 0.0000200
   g) 4346
   h) 34.86

11. Express the following numbers in decimal notation:

   a) 3.45×10^{7}
   b) $3,808 \times 10^{-4}$
   c) 2.3×10^{5}
   d) 4.000×10^{-5}

   e) 3×10^{3}
   f) 5.64×10^{6}
   g) 2.0×10^{1}
   h) 4×10^{1}

12. Carry out the following multiplications, expressing each answer in scientific notation. Be sure each answer contains the correct number of significant figures and that it is justified.

   a) $(4.0 \times 10^{3}) \times (3.0 \times 10^{5})$
   b) $(3.0 \times 10^{-5}) \times (5.0 \times 10^{-6})$
   c) $(2.0 \times 10^{-11}) \times (8.0 \times 10^{7})$
   d) $(4.0 \times 10^{8}) \times (4.0 \times 10^{-5})$
   e) $(3.74 \times 10^{3}) \times (2.894 \times 10^{3})$
   f) $(4.00 \times 10^{2}) \times (3.0 \times 10^{-5})$
   g) $(3.8 \times 10^{-4}) \times (3.74 \times 10^{-5}) \times (2.022 \times 10^{9})$
   h) $(7.00 \times 10^{9}) \times (2.034 \times 10^{-3}) \times (2.222 \times 10^{-4})$

13. Carry out the following divisions, expressing each answer in scientific notation. Be sure each answer contains the correct number of significant figures and that it is

justified.

a) $\dfrac{5.0 \times 10^5}{3.0 \times 10^3}$

b) $\dfrac{5.0 \times 10^5}{3.0 \times 10^{-3}}$

c) $\dfrac{5.0 \times 10^{-3}}{3.0 \times 10^5}$

d) $\dfrac{4.0 \times 10^{-3}}{3.0 \times 10^{-5}}$

e) $\dfrac{9.84 \times 10^8}{3.74 \times 10^7}$

f) $\dfrac{2.0 \times 10^{-8}}{1.62 \times 10^3}$

g) $\dfrac{6 \times 10^5}{2 \times 10^{-3}}$

h) $\dfrac{5.126 \times 10^{-4}}{2.0 \times 10^{-3}}$

14. Perform the following mathematical operations, expressing each answer in scientific notation. Be sure each answer contains the correct number of significant figures and that it is justified.

a) $\dfrac{(5.0 \times 10^1) \times (7.0 \times 10^5)}{2.0 \times 10^6}$

b) $\dfrac{(7.43 \times 10^{-3}) \times (6.54 \times 10^{-4})}{3.0 \times 10^{-11}}$

c) $\dfrac{(8.56 \times 10^{-4}) \times (2.3 \times 10^5)}{(5.445 \times 10^{-6}) \times (4.65 \times 10^5)}$

d) $\dfrac{(2.00 \times 10^3) \times (4.4 \times 10^2)}{(6.2 \times 10^5) \times (8.4 \times 10^2)}$

15. Perform the following mathematical operations, expressing each answer in scientific notation. Be sure your answers contain the correct number of significant figures and that they are justified.

a) $(3.772 \times 10^4) + (4.244 \times 10^4)$

b) $(4.04 \times 10^4) - (2.3 \times 10^4)$

c) $(5.0 \times 10^4) - (5.0 \times 10^3)$

d) $(4.78 \times 10^5) + (4.78 \times 10^6)$

e) $(3.0 \times 10^7) - (2.000 \times 10^5)$

f) $(7.00 \times 10^7) + (6.0 \times 10^6) + (6 \times 10^5)$

g) $(4.1 \times 10^6) - (3.78 \times 10^2)$

h) $(4.2222 \times 10^6) + (4.2 \times 10^3)$

MULTIPLE–CHOICE EXERCISES

1. If the accepted value for a measurement is 3.54 inches, which of the following exhibits the best precision?

    a) 3.95, 3.64, 2.85
    b) 3.54, 3.53, 3.55
    c) 3.42, 3.51, 3.50
    d) 2.85, 3.54, 5.64

2. Which of the following numbers contains 3 significant figures?

    a) 0.00333
    b) 0.0003330
    c) 33,300.0
    d) 3.3300

3. How many nonsignificant zeros are there in the measured number 0.03070?

    a) 1
    b) 2
    c) 3
    d) 4

4. The number 52,555, when rounded off to 3 significant figures, would appear as:

    a) 525
    b) 526
    c) 52500
    d) 52600

5. The number 0.030450, when rounded off to 3 significant figures, would appear as:

    a) 0.03045
    b) 0.0304
    c) 0.0305
    d) 0.031

6. Which of the following underlined numbers does not contain an unlimited number of significant figures?

    a) 14 lemons
    b) 1 foot = 12 inches
    c) 16 gallons = volume of container
    d) 4 quarts = 1 gallon

7. The number 0.0030 expressed in scientific notation becomes:

   a) 3.0×10^{-3}
   b) 3.0×10^{3}
   c) 3×10^{-3}
   d) 0.003×10^{-3}

8. Which of the following is the decimal equivalent of 7.5×10^{5}?

   a) 750,000
   b) 7,500,000
   c) 0.0000075
   d) 0.7500000

9. Which of the following numbers is smallest in magnitude?

   a) 3.02×10^{4}
   b) 3.02×10^{-4}
   c) 3.02×10^{-5}
   d) 3.20×10^{-5}

10. Which one of the following mathematical expressions is evaluated correctly?

   a) $10^{6} \times 10^{2} = 10^{12}$
   b) $\dfrac{10^{6}}{10^{2}} = 10^{3}$
   c) $\dfrac{10^{-6}}{10^{-2}} = 10^{-4}$
   d) $\dfrac{10^{6}}{10^{-2}} = 10^{4}$

11. The number 212.53×10^{4}, when justified, becomes:

   a) 21253×10^{6}
   b) 21253×10^{2}
   c) 2.1253×10^{6}
   d) 2.1253×10^{2}

12. The number 11.3×10^{-4}, when justified, becomes:

   a) 113×10^{-5}

   b) 113×10^{-3}

   c) 1.13×10^{-5}

   d) 1.13×10^{-3}

13. The answer obtained by multiplying (3.00×10^2) by (2.00×10^4) is:

   a) 6×10^6

   b) 6×10^8

   c) 6.00×10^6

   d) 6.00×10^8

14. The answer obtained by dividing (8.00×10^6) by (2.0×10^4) is:

   a) 4.0×10^{10}

   b) 4.0×10^2

   c) 4.00×10^{10}

   d) 4.00×10^2

15. The answer obtained by adding the measurements 71.2 and 31.210 is:

   a) 102

   b) 102.4

   c) 102.41

   d) 102.410

In questions 16–20 select the answer which is correctly justified and contains the correct number of significant figures.

16. $(9 . 342 \times 10^4) \times (3.9 \times 10^{-4}) =$

   a) 36

   b) 36.4

   c) 3.64×10^9

   d) 3.64×10^0

17. $\dfrac{7.54 \times 10^{-3}}{9.2 \times 10^{-3}} =$

    a) 8.2×10^{-1}

    b) 8.2×10^{-7}

    c) 0.82×10^{0}

    d) 0.81956522

18. $(7.16 \times 10^{3}) + (8.17 \times 10^{4}) =$

    a) 8.886×10^{3}

    b) 8.886×10^{4}

    c) 8.89×10^{3}

    d) 8.89×10^{4}

19. $\dfrac{(2.50 \times 10^{-3})}{(4.43 \times 10^{2}) \times (3.5 \times 10^{-4})} =$

    a) 2.0×10^{-9}

    b) 1.6×10^{-2}

    c) 1.61×10^{-9}

    d) 0.16×10^{-1}

20. $\dfrac{(4.0 \times 10^{2}) \times (5.0 \times 10^{3})}{(2.0 \times 10^{4}) \times (3.0 \times 10^{5})} =$

    a) 3.3×10^{-4}

    b) 3×10^{-4}

    c) 3.0×10^{6}

    d) 33×10^{-5}

ANSWERS TO PROBLEMS SET

1.   Student A has good precision and good accuracy.
       Student B has poor precision and good accuracy.
       Student C has good precision and poor accuracy.

2.   a) 2   c) 3   e) 4   g) 5
       b) 3   d) 2   f) 4   h) 1

3. a) infinite, 144 is a defined number.
    b) infinite, results from counting.
    c) infinite, 32 is defined.
    d) 3

4. a) none   d) 1
    b) 2   e) 5
    c) none

5. a) 6460   e) 985
    b) 1.48   f) 16.0
    c) 4.90   g) 0.0301
    d) 64.2   h) 0.104

6. a) 12   e) 20
    b) 6.0   f) 30
    c) 17.830   g) 44000
    d) 17   h) 0.024

7. a) 1740   e) 950.00
    b) 1700   f) 0.07
    c) 4   g) 0.7
    d) 1.98   h) 200

8. a) 12
    b) 1.1
    c) 0.588
    d) 230,000

9. a) 21   e) 888.9
    b) 42.0   f) 2344
    c) 0.1   g) 0.1175
    d) 0.663   h) 60.0

10. a) 3.45×10^{2} e) 3.40×10^{-1}

    b) 4.28×10^{-3} f) 2.00×10^{-5}

    c) 4.04003×10^{1} g) 4.346×10^{3}

    d) 2×10^{3} h) 3.486×10^{1}

11. a) 34,500,000   e) 0.003
    b) 0.0003808   f) 5,640,000
    c) 230,000   g) $2\overline{0}$
    d) 0.00004000   h) 40

12. a) 1.2×10^{9} e) 1.08×10^{7}

    b) 1.5×10^{-10} f) 1.2×10^{-2}

    c) 1.6×10^{-3} g) 2.9×10^{1}

    d) 1.6×10^{4} h) 3.16×10^{3}

13. a) 1.7×10^2 e) 2.63×10^1

     b) 1.7×10^8 f) 1.2×10^{-11}

     c) 1.7×10^{-8} g) 3×10^8

     d) 1.3×10^2 h) 2.6×10^{-1}

14. a) 1.8×10^1

     b) 1.6×10^5

     c) 7.8×10^1

     d) 1.7×10^{-3}

15. a) 8.016×10^4 e) 3.0×10^7

     b) 1.7×10^4 f) 7.66×10^7

     c) 4.5×10^4 g) 4.1×10^6

     d) 5.26×10^6 h) 4.2264×10^6

ANSWERS TO MULTIPLE CHOICE EXERCISES

1.	b	11.	c
2.	a	12.	d
3.	b	13.	c
4.	d	14.	b
5.	b	15.	b
6.	c	16.	a
7.	a	17.	a
8.	a	18.	d
9.	c	19.	b
10.	c	20.	a

CHAPTER 3
Unit Systems
and Dimensional Analysis

REVIEW OF CHAPTERS OBJECTIVES

1. <u>List the basic metric system units and the prefixes used to indicate multiple and subunits of the basic unit</u> (section 3.1).

The metric system is the system of units of measure used in scientific work. There are three basic metric system units.

The <u>meter</u> is the basic unit of length in the metric system.

The <u>gram</u> is the basic unit of mass in the metric system.

The <u>liter</u> is the basic unit of volume in the metric system.

These basic units are then multiplied by powers of ten to form smaller or larger units. The names of the larger or smaller units are constructed from the basic unit name by attaching a prefix which tells the power of ten involved.

Here are the prefixes and the power of ten which they indicate:

tera-	10^{12}
giga-	10^{9}
mega-	10^{6}
kilo-	10^{3}
hecto-	10^{2}
deca-	10^{1}
deci-	10^{-1}
centi-	10^{-2}

milli-	10^{-3}
micro-	10^{-6}
nano-	10^{-9}
pico-	10^{-12}

The meaning of a prefix always means the same thing regardless of the basic unit it modifies. For example, a kilometer is a thousand meters and a kiloliter is a thousand liters.

2. <u>Explain the difference between the mass and weight of an object</u> (section 3.1).

The terms mass and weight are frequently used interchangeably. In most cases this does no harm, but technically it is incorrect. Mass and weight refer to different properties of matter and their different meanings should be understood.

Mass is a measure of the total quantity of matter in an object, while weight is a measure of the gravitational attraction for that body.

The mass of a substance is constant, while the weight of an object is variable depending on the distance between the centers of the two objects. This is why it is not correct to use mass and weight interchangeably.

As an example of the difference in meaning of mass and weight, consider an astronaut. In outer space, we say he is weightless since he is outside the earth's gravitational attraction, but he still contains the same quantity of matter (mass) as on earth.

3. <u>Set up and work unit system conversion problems using dimensional analysis</u> (section 3.2).

Many times we need to change from one unit of measure to a different unit of measure. The method used to make this conversion is called dimensional analysis. This is the most powerful tool you can learn for problem solving. You should practice using it until you become very proficient. Remembering that all numbers resulting from a measurement or counting have a unit which goes along with them and always writing these units down will keep you from making many mistakes.

The following steps are used to solve problems using dimensional analysis:

1. Identify the known or given quantity (both a numerical value and units) and the units of the new quantity to be determined.

2. Multiply the given quantity by one or more conversion factors in a manner such that the unwanted (original) units are cancelled out leaving only the new desired units.

3. Perform the mathematical operations indicated by the conversion factor "set-up".

Example:

1. How many seconds are contained in 4.0 minutes?

   Step 1: The given quantity is 4.0 minutes. The unit of the desired quantity is seconds.

   4.0 minutes = ? seconds

   Step 2: The equality needed to derive the conversion factor is one relating minutes to seconds.

   60 seconds = 1 minute

   The conversion factor can be written in two forms.

   $$\frac{60 \text{ seconds}}{1 \text{ minute}} \quad \text{and} \quad \frac{1 \text{ minute}}{60 \text{ seconds}}$$

   Of these two factors, the $\frac{60 \text{ seconds}}{1 \text{ minute}}$ will be used. It allows the minutes to be cancelled leaving seconds, the desired new unit.

   $$4.0 \text{ minutes} \times \frac{60 \text{ seconds}}{1 \text{ minute}} = ? \text{ seconds}$$

   For a unit to be cancelled, it must appear in both the numerator and the denominator. In this example, the given quantity was 4.0 minutes so the conversion factor used had to have minutes in the denominator.

   Step 3: Multiply 4.0 times 60 to obtain the answer.

   $$(4.0 \times 60) \text{ seconds} = 240 \text{ seconds (calculator and correct answer)}$$

Since the conversion factor is derived from a definition, it contains an unlimited number of significant figures and will not limit in any way the number of significant figures in the answer. The answer contains 2 significant figures since the given quantity, 4.0 minutes, contains 2 significant figures.

2. How many yards are contained in 108 inches?

Step 1: The given quantity is 108 inches. The unit of the desired quantity is yards.

108 inches = ? yards

Step 2: The equality needed to derive the conversion factor is one relating yards to inches.

36 inches = 1 yard

The conversion factor can be written in 2 forms:

$$\frac{36 \text{ inches}}{1 \text{ yard}} \quad \text{and} \quad \frac{1 \text{ yard}}{36 \text{ inches}}$$

Of these 2 factors, the $\frac{1 \text{ yard}}{36 \text{ inches}}$ will be used. It allows the inches to be cancelled leaving yard, the desired new unit.

$$108 \text{ } \cancel{\text{inches}} \times \left(\frac{1 \text{ yard}}{36 \text{ } \cancel{\text{inches}}} \right) = ? \text{ yards}$$

Step 3: All that is left to do in step 3 is to combine numerical terms to get a final answer.

$$\frac{108}{36} \text{ yards} = 3 \text{ yards (calculator answer)}$$
$$= 3.00 \text{ yards (correct answer)}$$

The correct answer is 3.00 yards since 108 contains 3 significant figures and the answer must also contain the same number. The conversion factor contains an unlimited number of significant figures and has no effect on the answer.

Notice this example involved a division while the previous one did not. This is the power of dimensional analysis since it tells you when to multiply or divide. The 36 inches was placed in the denominator so the inches would cancel which

resulted in the division. The only way there could
have been a multiplication would be to use $\dfrac{36 \text{ yard}}{1 \text{ inch}}$

which is an inappropriate conversion factor for this
problem.

If more than one conversion factor is needed the
process is the same only more steps are involved.

4. <u>Construct two conversion factors from a given equivalence
between two quantities</u> (section 3.2).

Conversion factors are constructed from two quantities
which are equal. For example, 4 quarts equal one gallon.
This can also be stated in the mathematical expression

   1 gallon = 4 quarts

Dividing both sides of the equation by one gallon gives:

   $\dfrac{1 \text{ gallon}}{1 \text{ gallon}} = \dfrac{4 \text{ quarts}}{1 \text{ gallon}}$

One gallon divided by one gallon is one which gives

   $1 = \dfrac{4 \text{ quarts}}{1 \text{ gallon}}$

The fraction is the conversion factor and is read as 4
quarts per one gallon.

Dividing both sides of the equation by 4 quarts gives

   $\dfrac{1 \text{ gallon}}{4 \text{ quarts}} = \dfrac{4 \text{ quarts}}{4 \text{ quarts}}$

Four quarts divided by four quarts is one which gives

   $\dfrac{1 \text{ gallon}}{4 \text{ quarts}} = 1$

The fraction is the conversion factor and is read as 1
gallon per 4 quarts.

These two conversion factors are reciprocals of each
other and result from the same equality. Two conversion
factors are always obtainable from one equality. Notice
that the 4 always remains with the same units (quarts).

Another way of obtaining the second conversion factor is
to write the reciprocal of the first.

For example, a car can travel 60 miles per one hour

$$\frac{60 \text{ miles}}{1 \text{ hour}}$$

or, in one hour a car could travel 60 miles

$$\frac{1 \text{ hour}}{60 \text{ miles}}$$

All that was done was the fraction was inverted. Notice that the 60 always remains with the miles.

5. **Distinguish between density and specific gravity, including the units of each** (section 3.3).

Density is the ratio of the mass of an object to the volume occupied by that object. The mathematical equation for density is:

$$\text{density} = \frac{\text{mass}}{\text{volume}}$$

The most frequently used units for density are grams per cubic centimeter (g/cm^3) for solids, grams per milliliter (g/mL) for liquids, and grams per liter (g/L) for gases.

Specific gravity of a solid or liquid is the ratio of the density of that substance to the density of water at 4° C. For gases, specific gravity involves a density comparison with air instead of water.

In calculating the specific gravity of a substance, both densities must be expressed in the same units, and as a result specific gravity is unitless since the identical sets of density units cancel.

Example:

A solid object has a mass of 4.17 g and a volume of 2.02 cm^3. What is its density and specific gravity?

$$\text{density} = \frac{\text{mass}}{\text{volume}} = \frac{4.17 \text{ g}}{2.02 \text{ cm}^3} = 2.0643564 \text{ g/cm}^3 \text{ (calculator answer)}$$

$$= \underline{2.06 \text{ g}} \text{ (correct answer)}$$
$$cm^3$$

$$\text{specific gravity} = \frac{\text{density of object}}{\text{density of water at } 4^{\circ} \text{ C}} =$$

$$= \frac{2.06 \ \cancel{g/cm^3}}{1.000 \ \cancel{g/cm^3}} = 2.06 \ \text{(calculator and correct answer)}$$

6. <u>Calculate a substance's density given its mass and volume and use density as a conversion factor between mass and volume or vice versa</u> (section 3.3).

Examples:

1. Calculate the density of aluminum if 26.7 grams occupies a volume of 9.89 cm^3.

   Solution:

   Using the formula, density = $\dfrac{\text{mass}}{\text{volume}}$, we have

   density = $\dfrac{26.7 \ g}{9.89 \ cm^3}$ = 2.6996967 g/cm^3 (calculator answer)

   = 2.70 g/cm^3 (correct answer)

2. Given that the density of mercury is 13.6 g/mL, what volume, in mL, will 253 g of mercury occupy?

   Solution: Apply the three steps of dimensional analysis.

   Step 1: The given quantity is 253 g. The unit of the desired quantity is mL.

   253 g = ? mL

   Step 2: The conversion factor is the density which can be written as $\dfrac{13.6 \ g}{1 \ mL}$ and $\dfrac{1 \ mL}{13.6 \ g}$.
   Since g are to be converted to mL, use the fraction on the right so the g units will cancel.

   $253 \ \cancel{g} \times \left(\dfrac{1 \ mL}{13.6 \ \cancel{g}} \right) = ? \ mL$

   Step 3: Performing the indicated mathematical operations gives:

   $\left(\dfrac{253 \times 1}{13.6} \right) mL$ = 18.602941 mL (calculator answer)

   = 18.6 mL (correct answer)

3. What is the mass in grams of a 255 mL sample of sulfuric acid with a density of 1.30 g/mL?

Solution:

Step 1: The given quantity is 255 mL. The unit of desired quantity is g.

255 mL = ? g

Step 2: The conversion factor is the density which can be written as $\frac{1.30 \text{ g}}{1 \text{ mL}}$ and $\frac{1 \text{ mL}}{1.30 \text{ g}}$.

Since mL are to be converted to g, use the fraction on the left so the mL units will cancel.

$$255 \, \cancel{mL} \times \left(\frac{1.30 \text{ g}}{1 \, \cancel{mL}} \right) = ? \text{ g}$$

Step 3: Performing the indicated mathematical operations gives:

$$\frac{255 \times 1.30}{1} \text{ g} = 331.5 \text{ g (calculator answer)}$$
$$= 332 \text{ g (correct answer)}$$

7. <u>Set up and solve, using dimensional analysis, problems involving percentages, treating percentage as a conversion factor</u> (section 3.4).

Percent means parts per 100 parts -- that is, it is the number of specific items in a group of 100 items. The quantity 50% means 50 items per 100 total items.

The difficulty in writing conversion factors from percentages is determining the correct units to be used. This difficulty can be avoided if it is remembered to consider the complete context of the problem to determine the complete appropriate unit.

Example: A sample of iron ore was found to be 42.8% iron. How many grams of ore are needed to produce 382 grams of iron?

Solution:

Step 1: The given quantity is 382 grams of iron. The unit of the desired quantity is grams of iron ore.

382 grams of iron = ? grams of iron ore

Step 2: Since the units of mass on the given quantity are grams, the units of mass used in the conversion factor must be grams so they will cancel.

$$\frac{42.8 \text{ grams}}{100 \text{ grams}}$$

But grams are not the complete units. The context of the problem states there are 42.8 parts of iron per 100 parts of iron ore, and this must also be included in the units of the conversion factor to obtain:

$$\frac{42.8 \text{ grams of iron}}{100 \text{ grams of iron ore}}$$

The reciprocal of this can also be written as:

$$\frac{100 \text{ grams of iron ore}}{42.8 \text{ grams of iron}}$$

and it is this latter factor that is used to cancel grams of iron, the given quantity.

$$382 \text{ } \cancel{\text{grams of iron}} \times \frac{100 \text{ grams of iron ore}}{42.8 \text{ } \cancel{\text{grams of iron}}}$$

$$= ? \text{ grams of iron ore}$$

Notice that the complete unit must be cancelled and not just grams and the units of the answer are grams of iron ore, not just grams.

Step 3: Performing the indicated mathematical operations gives:

$$\left(\frac{382 \times 100}{48.2}\right) \text{ grams iron ore}$$

$$= 792.53112 \text{ grams of iron ore}$$
(calculator answer)
$$= 793 \text{ grams of iron ore}$$
(correct answer)

8. <u>Know and be able to use the interrelationships between Fahrenheit, Celsius, and Kelvin temperature scales</u> (section 3.5).

The relationships between the temperature scales are given by the following equations:

$$°F = 9/5(°C) + 32$$

$$°C = 5/9(°F - 32)$$

$$K = °C + 273$$

$$°C = K - 273$$

Fahrenheit temperatures cannot be converted directly to Kelvin, but must first be converted to Celsius; Kelvin cannot be converted directly to Fahrenheit, but must first be converted to Celsius.

Example:

1. Room temperature is 72 °F. What is this temperature on: (a) the Celsius scale and (b) Kelvin scale?

   Solution:

   a) Substituting into the appropriate equation

   $$°C = 5/9(°F - 32), \text{ we get}$$

   $$°C = 5/9(72 - 32) = 5/9(40)$$
   $$= 22.222222 \text{ °C (calculator answer)}$$
   $$= 22 \text{ °C (correct answer)}$$

   A common mistake in converting from Fahrenheit to Celsius is not subtracting before multiplying by 5/9.

   b) As stated before, Fahrenheit temperatures cannot be directly converted to the Kelvin scale. The conversion done in part (a) must first be done and then the result converted to Kelvin using the equation K = °C + 273.

   $$K = 22 \text{ °C} + 273 = 295 \text{ K}$$

2. Water is most dense at 4 °C. What is this temperature on the Fahrenheit scale?

   Solution:

   Substituting into the appropriate equation,

$$^{\circ}F = 9/5(^{\circ}C) + 32, \text{ we get}$$

$$^{\circ}F = 9/5(4) + 32 = 7.2 + 32 = 39.2 \ ^{\circ}F$$

(calculator answer)

$$= 39 \ ^{\circ}F \text{ (correct answer)}$$

PROBLEM SET

1. The basic unit of length in the metric system is the
   _____.

2. The liter is the basic unit of _____ in the metric
   system.

3. The basic unit of mass in the metric system is the
   _____.

4. What are the powers of ten associated with the following
   metric system prefixes?

   a) kilo   e) micro
   b) mega   f) pico
   c) centi   g) deca
   d) milli   h) nano

5. Using the dimensional analysis method of problem solving,
   carry out the following metric-metric conversions:

   a) 35.2 kilograms = ? grams
   b) 334.6 milligrams = ? grams
   c) 25.4 centimeters = ? meters
   d) 0.234 centimeters = ? millimeters
   e) 224.0 milliliters = ? liters
   f) 11.2 liters = ? milliliters

6. Using the dimensional analysis method of problem solving,
   carry out the following metric-English conversions:

   a) 32.8 grams = ? ounces
   b) 39.37 millimeters = ? inches
   c) 22.4 liters = ? quarts
   d) 427 grams = ? pounds
   e) 528 centimeters = ? feet
   f) 382 milliliters = ? gallons

7. Using the dimensional analysis method of problem solving, carry out the following English-metric conversions:

a) 50.0 pounds = ? kilograms
b) 16.0 fluid ounces = ? milliliters
c) 7.054 ounces = ? grams
d) 1.5 feet = ? centimeters
e) 24.8 gallons = ? liters
f) 32.4 yards = ? meters

8. A sample of ore is found to have a mass of 17.3 grams and a volume of 24.8 cm^3. What is its density?

9. What is the volume occupied by 15.3 grams of a liquid which has a density of 0.536 g/mL?

10. A solid substance has a density of 1.85 g/cm^3. What volume, in cm^3, will 25 g of the substance occupy?

11. Gasoline has a density of 0.56 g/mL. What would be the mass of 16 liters of gasoline?

12. What is the mass of a piece of copper which has a volume of 2.45 cm^3 if the density of copper is 8.93 g/cm^3?

13. If a wine contains 15.0 percent alcohol, how many milliliters of alcohol are there in 2.0 liters of the wine?

14. If a solution contains 35.0 percent water, how many grams of water are there in 452 grams of the solution?

15. Do the following temperature scale conversions:

a) 52 $^\circ$C to $^\circ$F
b) 138 $^\circ$C to K
c) 532 K to $^\circ$F

d) 43 K to $^\circ$C
e) 98 $^\circ$F to $^\circ$C
f) 212 $^\circ$F to K

MULTIPLE CHOICE EXERCISES

1. In the metric system, the fundamental units of mass, volume and length are, respectively:

a) liter, gram, meter
b) meter, liter, gram
c) gram, liter, meter
d) gram, meter, liter

2. In which of the following is the metric system prefix incorrectly paired with a power of ten?

   a) kilo- and 10^3

   b) milli- and 10^{-6}

   c) centi- and 10^{-2}

   d) mega- and 10^6

3. Which of the following statements is incorrect?

   a) a microgram is smaller than a milligram
   b) a kilometer is larger than a decameter
   c) a megaliter is smaller than a centiliter
   d) a decigram is larger than a nanogram

4. A distance of one centimeter relates most closely to which of the following?

   a) diameter of the wire in a paper clip
   b) thickness of your little finger
   c) diameter of a half-dollar coin
   d) height of a 2-year-old child

5. A mass of one kilogram relates most closely to which of the following?

   a) mass of a nickel coin
   b) mass of 6 thumbtacks
   c) mass of a cantaloupe
   d) mass of a large football player

6. Which of the following conversion factor unit "set-ups" would take you from kilograms to milligrams:

   a) kg x $\dfrac{cg}{kg}$ x $\dfrac{cg}{mg}$

   b) kg x $\dfrac{kg}{g}$ x $\dfrac{g}{mg}$

   c) kg x $\dfrac{g}{kg}$ x $\dfrac{mg}{g}$

   d) kg x $\dfrac{cg}{kg}$ x $\dfrac{g}{cg}$

7. Which of the following conversion factor unit "set-ups" would be appropriate to change $\frac{km}{hr}$ to $\frac{ft}{min}$?

a) $\frac{km}{hr}$ x $\frac{hr}{min}$ x $\frac{m}{ft}$

b) $\frac{km}{hr}$ x $\frac{m}{km}$ x $\frac{ft}{m}$ x $\frac{min}{hr}$

c) $\frac{km}{hr}$ x $\frac{km}{ft}$ x $\frac{hr}{min}$

d) $\frac{km}{hr}$ x $\frac{hr}{min}$ x $\frac{m}{km}$ x $\frac{ft}{m}$

8. In which of the following conversion factors do both numbers contain an unlimited number of significant figures?

a) $\frac{2.540 \text{ cm}}{1 \text{ inch}}$

b) $\frac{12 \text{ inches}}{1 \text{ foot}}$

c) $\frac{454 \text{ grams}}{1 \text{ pound}}$

d) $\frac{1 \text{ meter}}{39.37 \text{ in}}$

9. According to dimensional analysis, which of the following is the correct set-up for the problem "How many centimeters are there in 25 kilometers"?

a) 25 km x $\frac{1 \text{ m}}{10^3 \text{ km}}$ x $\frac{10^{-2} \text{ cm}}{1 \text{ m}}$

b) 25 km x $\frac{10^3 \text{ km}}{1 \text{ m}}$ x $\frac{1 \text{ m}}{10^{-2} \text{ cm}}$

c) 25 km x $\frac{10^3 \text{ m}}{1 \text{ km}}$ x $\frac{1 \text{ cm}}{10^{-2} \text{ m}}$

d) 25 km x $\frac{1 \text{ km}}{10^3 \text{ m}}$ x $\frac{10^{-2} \text{ m}}{1 \text{ cm}}$

10. How many pounds are contained in 452 milligrams?

    a) 205 pounds
    b) 0.996 pounds
    c) 9.96×10^{-4} pounds
    d) 1.00×10^{-3} pounds

11. How many pints are there in 5.00 liters?

    a) 2.36 pints
    b) 2.65 pints
    c) 9.43 pints
    d) 10.6 pints

12. Which of the following is not a density expression?

    a) 3.02 g/cm^3

    b) 0.303 mg/qt

    c) 2.33 cm/gal

    d) 3.01 lbs/mL

13. If 6 mL of A weighs 3 grams, the density of A is:

    a) 0.5 g/mL
    b) 3.0 g/mL
    c) 9.0 g/mL
    d) 18 g/mL

14. Mercury has a density of 13.6 g/mL. What volume, in mL, will 454 grams of mercury occupy?

    a) 6.17×10^{3} mL

    b) 33.4 mL

    c) 0.300×10^{-2} mL

    d) 617 mL

15. Which of the following is a correct interpretation of the fact that the specific gravity of a liquid is 2.3?

    a) 1 mL of the liquid weighs 2.3 g
    b) 2.3 mL of the liquid weighs 1.0 g
    c) a given volume of the liquid has a mass 2.3 times that of an equal volume of water
    d) water has a density 2.3 times greater than that of the liquid

16. What mass, in grams, of a 63.0% salt water solution is needed to supply 40.0 g of salt?

    a) 63.5 g
    b) 100.0 g
    c) 40.0 g
    d) 25.2 g

17. Which of the following comparisons of the size of a degree on the major temperature scales is correct?

    a) a Fahrenheit degree is larger than a Celsius degree
    b) a Fahrenheit degree and Kelvin degree are equal in size
    c) a Celsius degree is larger than a Kelvin degree
    d) a Kelvin degree is larger than a Fahrenheit degree

18. The number of degrees between the freezing point and boiling point of water on the Kelvin, Fahrenheit and Celsius scale is, respectively:

    a) 180, 100, 100
    b) 100, 100, 100
    c) 100, 180, 100
    d) 100, 180, 200

19. A temperature of 298° on the Kelvin scale is equivalent to:

    a) 25 $^{\circ}$C
    b) 98 $^{\circ}$C
    c) 498 $^{\circ}$C
    d) 571 $^{\circ}$C

20. If the temperature is 10 $^{\circ}$C then the reading of the Fahrenheit scale would be

    a) 50°
    b) 38°
    c) 18°
    d) 5.5°

PROBLEM SET ANSWERS

1. meter

2. volume

3. gram

4. a) 10^3 e) 10^{-6}

   b) 10^6 f) 10^{-12}

   c) 10^{-2} g) 10^1

   d) 10^{-3} h) 10^{-9}

5. a) $35.2 \, \cancel{kg} \times \left(\dfrac{10^3 \, g}{1 \, \cancel{kg}} \right) = 35{,}200 \, g$ (calculator and correct answer)

   b) $334.6 \, \cancel{mg} \times \left(\dfrac{10^{-3} \, g}{1 \, \cancel{mg}} \right) = 0.3346 \, g$ (calculator and correct answer)

   c) $25.4 \, \cancel{cm} \times \left(\dfrac{10^{-2} \, m}{1 \, \cancel{cm}} \right) = 0.254 \, m$ (calculator and correct answer)

   d) $0.234 \, \cancel{cm} \times \left(\dfrac{10^{-2} \, \cancel{m}}{1 \, \cancel{cm}} \right) \times \left(\dfrac{1 \, mm}{10^{-3} \, \cancel{m}} \right) = 2.34 \, mm$ (calculator and correct answer)

   e) $224.0 \, \cancel{mL} \times \left(\dfrac{10^{-3} \, L}{1 \, \cancel{mL}} \right) = 0.2240 \, L$ (calculator and correct answer)

   f) $11.2 \, \cancel{L} \times \left(\dfrac{1 \, mL}{10^{-3} \, \cancel{L}} \right) = 11{,}200 \, mL$ (calculator and correct answer)

6. a) $32.8 \, \cancel{g} \times \left(\dfrac{1 \, \cancel{lb}}{454 \, \cancel{g}} \right) \times \left(\dfrac{16 \, oz}{1 \, \cancel{lb}} \right) = 1.1559471 \, oz$ (calculator answer)
$= 1.16 \, oz$ (correct answer)

   b) $39.37 \, \cancel{mm} \times \left(\dfrac{10^{-3} \, \cancel{m}}{1 \, \cancel{mm}} \right) \times \left(\dfrac{1 \, \cancel{cm}}{10^{-2} \, \cancel{m}} \right) \times \left(\dfrac{1 \, in}{2.540 \, \cancel{cm}} \right)$

                                    $= 1.55 \, in$ (calculator answer)
                                    $= 1.550 \, in$ (correct answer)

   c) $22.4 \, \cancel{L} \times \left(\dfrac{1 \, qt}{0.946 \, \cancel{L}} \right) = 23.678647 \, qt$ (calculator answer)
                                    $= 23.7 \, qt$ (correct answer)

d) $427 \cancel{g} \times \left(\dfrac{1 \text{ lb}}{454 \cancel{g}}\right)$ = 0.94052863 lb (calculator answer)

$\qquad\qquad\qquad\qquad$ = 0.941 lb (correct answer)

e) $528 \cancel{cm} \times \left(\dfrac{1 \cancel{in}}{2.54 \cancel{cm}}\right) \times \left(\dfrac{1 \text{ ft}}{12 \cancel{in}}\right)$ = 17.322835 ft

$\qquad\qquad\qquad\qquad\qquad\qquad$ (calculator answer)

$\qquad\qquad\qquad\qquad\qquad\qquad$ = 17.3 ft (correct answer)

f) $382 \text{ mL} \times \left(\dfrac{10^{-3} \text{ L}}{1 \text{ mL}}\right) \times \left(\dfrac{1 \text{ qt}}{0.946 \text{ L}}\right) \times \left(\dfrac{1 \text{ gal}}{4 \text{ qt}}\right)$

$\qquad\qquad\qquad\qquad$ = 0.10095137 gal (calculator answer)

$\qquad\qquad\qquad\qquad$ = 0.101 gal (correct answer)

7. a) $50.0 \cancel{lb} \times \left(\dfrac{454 \cancel{g}}{1 \cancel{lb}}\right) \times \left(\dfrac{1 \text{ kg}}{10^3 \cancel{g}}\right)$ = 22.7 kg (calculator and

$\qquad\qquad\qquad\qquad\qquad\qquad\qquad\qquad$ correct answer)

b) $16.0 \cancel{fl\text{-}oz} \times \left(\dfrac{1 \cancel{qt}}{32 \cancel{fl\text{-}oz}}\right) \times \left(\dfrac{0.946 \cancel{L}}{1 \cancel{qt}}\right) \times \left(\dfrac{1 \text{ mL}}{10^{-3} \cancel{L}}\right)$

$\qquad\qquad\qquad$ = 473 mL (calculator and correct answer)

c) $7.054 \cancel{oz} \times \left(\dfrac{1 \cancel{lb}}{16 \cancel{oz}}\right) \times \left(\dfrac{453.6 \text{ g}}{1 \cancel{lb}}\right)$ = 199.9809 g

$\qquad\qquad\qquad\qquad\qquad\qquad\qquad$ (calculator answer)

$\qquad\qquad\qquad\qquad\qquad\qquad\qquad$ = 200.0 g

$\qquad\qquad\qquad\qquad\qquad\qquad\qquad$ (correct answer)

d) $1.5 \cancel{ft} \times \left(\dfrac{12 \cancel{in}}{1 \cancel{ft}}\right) \times \left(\dfrac{2.54 \text{ cm}}{1 \cancel{in}}\right)$ = 45.72 cm

$\qquad\qquad\qquad\qquad\qquad\qquad\qquad$ (calculator answer)

$\qquad\qquad\qquad\qquad\qquad\qquad\qquad$ = 46 cm (correct answer)

e) $24.8 \cancel{gal} \times \left(\dfrac{4 \cancel{qt}}{1 \cancel{gal}}\right) \times \left(\dfrac{0.946 \text{ L}}{1 \cancel{qt}}\right)$ = 93.8432 L

$\qquad\qquad\qquad\qquad\qquad\qquad\qquad$ (calculator answer)

$\qquad\qquad\qquad\qquad\qquad\qquad\qquad$ = 93.8 L

$\qquad\qquad\qquad\qquad\qquad\qquad\qquad$ (correct answer)

f) $32.4 \cancel{yd} \times \left(\dfrac{36 \cancel{in}}{1 \cancel{yd}}\right) \times \left(\dfrac{2.54 \cancel{cm}}{1 \cancel{in}}\right) \times \left(\dfrac{10^{-2} \text{ m}}{1 \cancel{cm}}\right)$

$\qquad\qquad\qquad\qquad$ = 29.62656 m (calculator answer)

$\qquad\qquad\qquad\qquad$ = 29.6 m (correct answer)

8. density = $\dfrac{\text{mass}}{\text{volume}}$ = $\dfrac{17.3 \text{ g}}{24.8 \text{ cm}^3}$ = 0.69758065 g/cm^3

$\qquad\qquad\qquad\qquad\qquad\qquad$ (calculator answer)

$\qquad\qquad\qquad\qquad\qquad\qquad$ = 0.698 g/cm^3

$\qquad\qquad\qquad\qquad\qquad\qquad$ (correct answer)

9. $15.3 \cancel{g} \times \left(\dfrac{1 \text{ mL}}{0.536 \cancel{g}}\right)$ = 28.544776 mL (calculator answer)

$\qquad\qquad\qquad\qquad$ = 28.5 mL (correct answer)

10. $25 \, \cancel{g} \times \left(\dfrac{1 \ cm^3}{1.85 \ \cancel{g}}\right) = 13.513514 \ cm^3$ (calculator answer)

$= 14 \ cm^3$ (correct answer)

11. $16 \, \cancel{L} \times \left(\dfrac{1 \ \cancel{mL}}{10^{-3} \ \cancel{L}}\right) \times \left(\dfrac{0.56 \ g}{1 \ \cancel{mL}}\right) = 8960$ g (calculator answer)

$= 9\overline{0}00$ g (correct answer)

12. $2.45 \, \cancel{cm^3} \times \left(\dfrac{8.93 \ g}{1 \ \cancel{cm^3}}\right) = 21.8785$ g (calculator answer)

$= 21.9$ g (correct answer)

13. 2.0 L of wine $\times \left(\dfrac{15 \ L \ alcohol}{100 \ L \ wine}\right) \times \left(\dfrac{1000 \ mL \ alcohol}{1 \ L \ alcohol}\right)$

$= 300$ mL alcohol (calculator answer)

$= 3\overline{0}0$ mL alcohol (correct answer)

14. $452 \, \cancel{g \ of \ solution} \times \left(\dfrac{35.0 \ g \ water}{100 \ \cancel{g \ solution}}\right)$

$= 158.2$ g water (calculator answer)

$= 158$ g water (correct answer)

15. a) $^{\circ}F = 9/5(52) + 32 = 93.6 + 32 = 125.6$

(calculator answer)

$= 126$ (correct answer)

b) $K = 138 + 273 = 411$ (calculator and correct answer)

c) $^{\circ}C = 532 - 273 = 259$

$^{\circ}F = 9/5(259) + 32 = 466.2 + 32 = 498.2$

(calculator answer)

$= 498$ (correct answer)

d) $^{\circ}C = 43 - 273 = -230$ (calculator answer)

$= -23\overline{0}$ (correct answer)

e) $^{\circ}C = 5/9(98 - 32) = 5/9(66) = 36.6666667$

(calculator answer)

$= 37$ (correct answer)

f) $^{\circ}C = 5/9(212 - 32) = 5/9(180) = 100$

(calculator answer)

$= 1\overline{00}$ (correct answer)

$K = 1\overline{00} + 273 = 373$

ANSWERS TO MULTIPLE CHOICE EXERCISES

1.	c		11.	d
2.	b		12.	c
3.	c		13.	a
4.	b		14.	b
5.	c		15.	c
6.	c		16.	a
7.	d		17.	d
8.	b		18.	c
9.	c		19.	a
10.	c		20.	a

CHAPTER 4
Basic Concepts About Matter

REVIEW OF CHAPTER OBJECTIVES

1. <u>Understand what is meant by the term matter</u> (section 4.1).

   Matter is anything which occupies space and has mass. Another way of thinking of occupying space is having volume. Having mass can be thought of as weight, although it is not entirely correct as you learned in the previous chapter.

   Even though an empty room seems to contain no matter, it in fact does. The air which fills the room occupies it -- that is, has the volume of the room -- and its mass could also be determined.

   Everything we can see, feel, or smell is matter. Gases are more difficult to think of as matter since we cannot see them.

2. <u>Know the shape and volume characteristics of the three states of matter</u> (section 4.2).

   Matter is classified into three physical states -- solid, liquid or gas. All the matter in nature exists in one of these three physical states.

   The characteristics of the three states of matter are summarized in the following table:

Physical State	Shape	Volume
solid	definite	definite
liquid	indefinite	definite
gas	indefinite	indefinite

3. <u>Given a property of a substance, classify it as a physical or chemical property</u> (section 4.3).

PROPERTIES are the distinguishing characteristics of a substance used in its identification and description.

PHYSICAL PROPERTIES are properties observable without changing a substance into another substance.

Example:

1. Color
2. Odor
3. Taste
4. Size
5. Physical state (solid, liquid or gas)
6. Boiling point
7. Density
8. Melting point
9. Conductivity

CHEMICAL PROPERTIES are properties that matter exhibits as it undergoes changes in chemical composition.

Example:

1. Rusting of iron
2. Souring of milk
3. Burning of wood
4. Decomposition on heating of sugar
5. Nonflammability of water

4. <u>Classify the changes that occur in matter as physical or chemical</u> (section 4.4).

Changes in matter, like properties, may be classified as physical or chemical.
PHYSICAL CHANGES are processes which do not alter the basic nature (chemical composition) of the substance under consideration. A new substance is never formed as a result of a physical change. The most common type of physical change is a change in physical state. Changes in size, shape and state of subdivision are examples of physical changes which are not changes of state.

CHEMICAL CHANGES are processes which involve a change in the basic nature (chemical composition) of the substance. Such changes always involve a conversion of the material or materials into one or more new substances with distinctly different properties and composition than the starting materials.

Example:

When iron rusts, pure shiny iron metal combines with the oxygen of the air to form the new reddish-brown substance iron oxide, which is composed of both iron and oxygen.

Physical changes always accompany a chemical change, but a chemical change never accompanies a physical change.

5. <u>Know the terminology used to describe the various changes of state</u> (section 4.4).

    1. Condensation is the change from the gaseous state to the liquid or solid state.

    2. Evaporation is the change from the liquid state to the gaseous state.

    3. Melting is the change from the solid state to the liquid state.

    4. Freezing is the change from the liquid state to the solid state.

    5. Sublimation is the change directly from the solid state to the gaseous state. Most substances go from solid state to liquid state to gaseous state. Dry ice is a common substance which undergoes sublimation.

6. <u>Distinguish between the characteristics of pure substances and mixtures</u> (section 4.5).

All matter can be divided into two classes, pure substances and mixtures.

A <u>pure substance</u> is a form of matter which always has a definite and constant composition. A pure substance can be broken down into its component parts only by chemical methods. For example, water, a pure substance, can be broken down into hydrogen and oxygen by electrolysis, a chemical process.

A <u>mixture</u> is a physical combination or collection, with a variable composition, of two or more pure substances. Mixtures can be broken down into their component parts by physical methods. For example, a sand and salt mixture can be broken down into sand and salt using the physical property of solubility. Salt is soluble in water and sand is not. After the salt is dissolved, the sand can be isolated using the physical method of filtration. The dissolved salt will pass through the filter paper while the undissolved sand remains behind on the filter paper.

Composition is an important distinguishing feature between pure substances and mixtures. Pure substances always have a definite composition while mixtures always have a variable composition. Other distinguishing features are listed on page 78 of the text (in Figure 4-3).

7. Explain the major differences between heterogeneous and homogeneous mixtures (section 4.5).

   A heterogeneous mixture contains visibly different parts or phases, each with different properties.

   A homogeneous mixture contains only one phase with uniform properties throughout it.

   A sand and salt mixture is an example of a heterogenous mixture. You can see the white salt and the brown sand parts of the mixture.

   A water solution of salt is an example of a homogeneous mixture. You can only see the liquid water phase, but the solid salt is uniformly dispersed throughout the water.

8. Understand the differences between an element and a compound (section 4.6).

   There are two kinds of pure substances -- elements and compounds.

   An element is a pure substance that cannot be broken down into simpler substances by ordinary chemical means.

   A compound is a pure substance that can be broken down into two or more simpler substances using chemical means.

   Elements are the fundamental "building blocks" for all types of matter. Elements cannot be formed from simpler substances or broken down into simpler forms of matter. Chemical compounds are formed when two or more elements combine together. For example, sodium (a soft, shiny, very reactive metal) and chlorine (a greenish, choking, poisonous gas) are both elements. They combine together to form white crystals of the chemical compound called sodium chloride which we use in everyday life as common table salt.

9. Know general trends concerning the discovery and abundance of the elements (section 4.7).

Table 4-3 on page 81 of your textbook gives the time frame over which the 106 known elements have been discovered. From the table, it can be seen that the majority (72) have been discovered since 1800.

Table 4-4, also on page 81 in your textbook, gives the distribution and abundance of the elements from various viewpoints.

10. Write the names when given the symbols, or the symbols when given the names of the more common elements (section 4.8).

Each of the 106 elements has a unique name. In the early 1800's, chemists adopted the practice of also using chemical symbols for the elements.

Chemical symbols are abbreviations for the names of the elements. These chemical symbols are used more frequently in referring to the elements than are the names themselves.

You would do well to learn the symbols of the following more common elements. Learning them is a key to having a successful experience in studying chemistry.

One way to aid yourself in learning them is to make a set of cards putting the name on one side and the symbol on the other. Then go through the cards looking at the symbol and giving the name, or vice versa, until you have learned them all.

Some of the elements have one-letter symbols. If a symbol consists of a single letter it is capitalized.

Name	Symbol
boron	B
carbon	C
fluorine	F
hydrogen	H
iodine	I
nitrogen	N
oxygen	O
phosphorus	P
sulfur	S

Some of the elements have double-letter symbols. In all double-letter symbols, the first letter is capitalized but the second letter is not.

Name	Symbol
aluminum	Al
argon	Ar
barium	Ba
beryllium	Be
calcium	Ca
chlorine	Cl
cobalt	Co
chromium	Cr
helium	He
lithium	Li
magnesium	Mg
neon	Ne
nickel	Ni
silicon	Si
zinc	Zn

Nine of the common elements have symbols which have no relationship to their English language name. For these elements, the symbol is derived from the Latin name. The Latin name is given in the following table to help you in understanding the symbol, but you are not required to know it.

English name	Latin name	Symbol
copper	cuprum	Cu
gold	aurum	Au
iron	ferrum	Fe
lead	plumbum	Pb
mercury	hydrargyrum	Hg
potassium	kalium	K
silver	argentum	Ag
sodium	natrium	Na
tin	stannum	Sn

PROBLEM SET

1. Anything which occupies space and has mass is called _____.

2. Solids have _____ shape and _____ volume.

3. Gases have _____ shape and _____ volume.

4. Liquids have _____ shape and _____ volume.

5. Classify each of the following as a chemical or physical property:

   a) reacts with oxygen   d) burns in air
   b) melts at 200°C   e) less dense than water
   c) yellow in color   f) solid at room temperature

6. Classify each of the following as a chemical or physical change:

   a) melting of ice   d) condensation of steam
   b) burning of wood   e) digestion of food
   c) evaporation of gasoline   f) rusting of iron

7. Give the name of the change of state associated with each of the following:

   a) ice becomes liquid water
   b) liquid water becomes steam
   c) molten aluminum becomes a solid
   d) ice changes directly into water vapor
   e) steam turns into liquid water

8. Tell whether the following are characteristics of a mixture or a pure substance:

   a) only one substance present
   b) components can be separated using physical means
   c) definite and constant composition
   d) properties vary with composition
   e) properties are always the same under a given set of conditions
   f) variable composition
   g) physical combination of two or more substances

9. Tell whether the following are characteristics of a homogeneous or a heterogeneous mixture.

   a) only one phase present
   b) two or more phases present
   c) uniform properties throughout mixture

10. A pure substance which cannot be broken down into simpler substances is an _____.

11. A pure substance which can be broken down into simpler substances is a _____.

12. How many elements were discovered in each of the following time periods?

a) ancient - 1700   d) 1901 - date
b) 1701 - 1800   e) 1951 - date
c) 1801 - 1900

13. Which element is most abundant in each of the following seven categories?

a) universe   e) hydrosphere
b) earth (including core)   f) human body
c) earth crust   g) vegetation
d) atmosphere

14. What chemical elements are represented by the following symbols?

a) Al   f) P
b) B   g) S
c) Ca   h) Sn
d) K   i) Zn
e) Ni

15. What are the chemical symbols of the following elements?

a) silver   f) lead
b) gold   g) tin
c) copper   h) neon
d) iron   i) cobalt
e) mercury

MULTIPLE CHOICE EXERCISES

1. Which of the following is a property of both liquids and solids?

a) definite shape
b) indefinite shape
c) definite volume
d) indefinite volume

2. Which of the following is a chemical property?

a) melts at 20°C
b) red in color
c) burns in air
d) sublimes

3. In which of the following pairs of properties are both properties physical properties?

   a) good reflector of light, flammable
   b) blue in color, decomposes upon heating
   c) melts at 35°C, is very hard
   d) has a high density, reacts with chlorine

4. Which of the following is <u>not</u> a chemical change?

   a) burning gasoline
   b) freezing water
   c) purifying copper ore
   d) decomposing sugar

5. When a substance undergoes a physical change, which of the following is always true?

   a) it melts
   b) its chemical composition remains the same
   c) a new substance is produced
   d) it changes state

6. The process of evaporation is a

   a) chemical change
   b) physical property
   c) chemical property
   d) physical change

7. Which of the following terms does not involve the liquid state?

   a) evaporation
   b) melting
   c) sublimation
   d) freezing

8. A mixture must

   a) contain at least two phases
   b) be separable using chemical mean only
   c) be heterogeneous
   d) contain at least two substances

9. Within a mixture, a phase is a region in which

   a) both solid and liquid are present
   b) the properties are uniform
   c) three or less substances are present
   d) three or more substances are present

10. Pure substances are always:

    a) colorless, transparent and nonflammable
    b) uniform throughout in properties
    c) decomposable into simpler substances using physical
       means
    d) compounds

11. A pure substance A is found to change, upon heating,
    into two new substances, B and C. From this we may
    conclude that:

    a) A and B are both elements
    b) B and C are both compounds
    c) A is a compound, B and C may or may not be elements
    d) A is an element, B and C are compounds

12. The number of known elements is _____ the number of
    known compounds.

    a) about the same as
    b) millions of times greater than
    c) approximately double
    d) very small compared to

13. Which of the following statements concerning the known
    elements is true?

    a) not all of them are naturally occurring
    b) the last of the elements was identified in 1885
    c) although the elements do not occur in nature in
       exactly equal abundances, their abundances are very
       close to being equal
    d) 112 different elements are known at present

14. In which of the following is hydrogen the most abundant
    element (in atom percent)?

    a) universe
    b) earth (including its core)
    c) atmosphere
    d) earth's crust

15. In which of the following is oxygen the most abundant
    element (in atom percent)?

    a) atmosphere
    b) earth's crust
    c) human body
    d) hydrosphere

16. Which of the following elements has a symbol which contains two letters?

    a) boron
    b) iodine
    c) lithium
    d) sulfur

17. Which of the following elements has a symbol which contains one letter?

    a) calcium
    b) tin
    c) gold
    d) fluorine

18. Which of the following symbols is paired up with the wrong element name:

    a) Al – aluminum
    b) Si – silver
    c) Br – bromine
    d) Mg – magnesium

19. Which of the following elements has a symbol which starts with a letter not the first letter of the element's English name?

    a) chlorine
    b) beryllium
    c) sodium
    d) neon

20. The symbols for the elements phosphorus, potassium, copper and silicon are, respectively:

    a) P, Pt, Co, S
    b) K, P, Cp, Sn
    c) P, K, Cu, Si
    d) Ph, Po, Co, Si

ANSWERS TO PROBLEMS SET?

1. matter

2. definite, definite

3. indefinite, indefinite

4. indefinite, definite

5. a) chemical   d) chemical
    b) physical   e) physical
    c) physical   f) physical

6. a) physical   d) physical
    b) chemical   e) chemical
    c) physical   f) chemical

7. a) melting   d) sublimation
    b) evaporation   e) condensation
    c) freezing

8. a) pure substance   e) pure substance
    b) mixture   f) mixture
    c) pure substance   g) mixture
    d) mixture

9. a) homogeneous
    b) heterogeneous
    c) homogeneous

10. element

11. compound

12. a) 13   d) 24
    b) 21   e) 8
    c) 48

13. a) hydrogen   e) hydrogen
    b) oxygen   f) hydrogen
    c) oxygen   g) hydrogen
    d) nitrogen

14. a) aluminum   f) phosphorus
    b) boron   g) sulfur
    c) calcium   h) tin
    d) potassium   i) zinc
    e) nickel

15. a) Ag   f) Pb
    b) Au   g) Sn
    c) Cu   h) Ne
    d) Fe   i) Co
    e) Hg

ANSWERS TO SELF TEST

1.	c	11.	c
2.	c	12.	d
3.	c	13.	a
4.	b	14.	a
5.	b	15.	b
6.	d	16.	c
7.	c	17.	d
8.	d	18.	b
9.	b	19.	c
10.	b	20.	c

CHAPTER 5
The Atom and Its Structure

REVIEW OF CHAPTER OBJECTIVES

1. Explain current scientific thought concerning atoms as
    stated in the postulates of the atomic theory of matter
    (section 5.1).

    Postulates of the atomic theory of matter are as follows:

    1. All matter is made up of small particles called
        atoms, of which 106 different "types" are known with
        each "type" corresponding to atoms of a different
        element.

        The atoms of the 106 known different elements are the
        basic building blocks of all that exists in nature.

    2. All atoms of a given type are similar to one another
        and significantly different from all other types.

        All atoms of the same element have the same chemical
        and physical properties which are distinctly
        different from those of atoms of another element.

    3. The relative number and arrangement of different
        types of atoms contained in a pure substance
        determine its identity.

        The number, kind and arrangement of different kinds
        of atoms determines a compound's chemical and
        physical properties.

        Water is composed of 2 atoms of hydrogen and one
        atom of oxygen and has distinctly different
        properties than hydrogen peroxide, composed of 2
        atoms of hydrogen and 2 atoms of oxygen.

    4. Chemical change is a union, separation or
        rearrangement of atoms to give new substances.

When a chemical reaction (change) takes place, new compounds or elements are formed which no longer have the chemical or physical properties of the starting material. In the rusting of iron, a chemical change, iron atoms react with oxygen atoms to form the new compound iron oxide.

5. <u>Only whole atoms can participate in or result from any chemical change, since atoms are considered to be indestructible during such changes.</u>

Since an atom is the smallest particle of an element that can exist, only whole atoms can take part in or result from a chemical change.

2. <u>Understand the difference between homoatomic and heteroatomic molecules and identify the type of substance in which each would be found</u> (section 5.2).

<u>Homoatomic molecules</u> are molecules in which all atoms present are the same element. Only elements can exist as homoatomic molecules.

The most common type of homoatomic molecule is a diatomic molecule (contains only 2 atoms). Oxygen, hydrogen, nitrogen, chlorine, fluorine, bromine and iodine form homoatomic diatomic molecules. Phosphorus is a tetraatomic (4 atoms) homoatomic molecule. Sulfur is an octaatomic (8 atoms) homoatomic molecule.

<u>Heteroatomic molecules</u> are molecules in which two or more different kinds of atoms are present. All heteroatomic molecules must be compounds since more than one kind of atom is present.

The designations diatomic, triatomic, tetraatomic, etc., apply to heteroatomic molecules as well as homoatomic molecules. Hydrogen sulfide (H_2S), composed of two atoms of hydrogen and one atom of sulfur, is a heteroatomic triatomic molecule. Methane gas (CH_4), composed of one atom of carbon and four atoms of hydrogen, is a heteroatomic pentaatomic molecule.

3. <u>Interpret a correctly written formula in terms of the number of elements and the number of atoms present</u> (section 5.3).

A <u>chemical formula</u> is a notation that contains the symbols of the elements present in a compound and numerical subscripts which indicate the relative number of atoms of each element present.

Example:

Interpret each of the following formulas in terms of how many atoms of each element are present in one structural unit of the substance:

a) C_2H_6O b) $Ca_3(PO_4)_2$ c) $(NH_4)_2SO_4$

Solutions:

a) 3 elements are present -- carbon, hydrogen and oxygen. There are 2 atoms of carbon, 6 atoms of hydrogen, and 1 atom of oxygen present. The subscripts give the number of atoms of each different element present; when no subscript is given only one atom of that element is present.

b) 3 elements are present, indicated by their symbols in the formula -- calcium, phosphorus and oxygen. There are 3 atoms of calcium, 2 atoms of phosphorus and 8 atoms of oxygen present. When a group of atoms is enclosed in parentheses, the subscript following the last parenthesis multiplies all subscripts inside the parentheses. Phosphorus has no subscript, so it is 1, and 1 x 2 = 2. Oxygen has a subscript of 4, so 4 x 2 = 8.

c) 4 elements are present -- nitrogen, hydrogen, sulfur and oxygen. There are 2 atoms of nitrogen, 8 atoms of hydrogen, one atom of sulfur and four atoms of oxygen present.

4. <u>Name the three major subatomic particles that make up an atom, tell where each is located within the atom and indicate the electrical charge and relative mass associated with each particle</u> (section 5.4).

Subatomic particles are particles which are smaller than atoms, and are the building blocks from which all atoms are made.

Information about the subatomic particles is summarized in the following table:

Particle	Abbreviation	Relative Mass	Charge	Location within atom
Electron	e	1	−1	extranuclear region
Proton	p	1836	+1	nuclear region
Neutron	n	1839	0	nuclear region

5. Define atomic number, mass number, and know how to determine the number of protons, neutrons and electrons present in an atom given these two numbers (section 5.5).

The atomic number is equal to the number of protons in the nucleus of an atom.

The mass number is equal to the number of protons plus neutrons in the nucleus of the atom.

The following equations show the relationship between the subatomic particle make-up of an atom and atomic number and mass number.

Number of protons = atomic number
Number of electrons = atomic number
Number of neutrons = (mass number) − (atomic number)

Example:

For an atom having an atomic number of 17 and a mass number of 37, determine:

a) the number of protons present
b) the number of neutrons present
c) the number of electrons present

Solution:

a) The number of protons is 17. The atomic number is always equal to the number of protons present.

b) The number of neutrons is 20. The number of neutrons is always obtained by subtracting the atomic number from the mass number.

    (protons + neutrons) − (protons) = neutrons
       mass number   atomic number

c) The number of electrons present is 17. In a neutral

atom, the number of electrons always equals the
number of protons (atomic number).

6. Define what isotopes are and be able to write the symbol
    for an isotope (section 5.6).

    Isotopes are atoms which have the same number of protons
    and electrons but different numbers of neutrons.
    Isotopes of the same element will always have the same
    atomic number (number of protons) and different mass
    numbers (number of protons + neutrons).

    Different isotopes of an element are distinguished from
    each other using the following notation:

    $$_Z^A\text{symbol}$$

    The atomic number whose general symbol is Z is written as
    a subscript to the left of the elemental symbol. The
    mass number whose general symbol is A is also written to
    the left of the elemental symbol, but as a superscript.

    Example:

        Use correct notation to distinguish between the three
        isotopes of uranium (U, atomic number 92) with mass
        numbers of 234, 235 and 238.

    Solution:

        The atomic number is 92, so Z = 92. The mass numbers
        are 234, 235 and 238, so A = 234, 235, 238. The
        elemental symbol is U. The three isotopes are:

        $$_{92}^{234}\text{U} \qquad _{92}^{235}\text{U} \qquad _{92}^{238}\text{U}$$

7. Calculate the atomic weight of an element from the
    isotopic masses and percentage abundances of its
    isotopes (section 5.7).

    Atomic weight is the relative mass of an average atom of
    an element on a scale using atoms of $_6^{12}\text{C}$ as the reference.

    Isotopic mass is the mass of a specific isotope of an
    element.

    Percentage abundance is the number of atoms of a
    particular isotope found per 100 atoms of the element.

Both isotopic masses and percentage abundances are experimentally determined numbers.

Example:

Chlorine occurs in nature in two isotopic forms -- $^{35}_{17}Cl$ and $^{37}_{17}Cl$. The isotopic mass of $^{35}_{17}Cl$ is 34.9689 amu and its percentage abundance is 75.53%. For $^{37}_{17}Cl$ the isotopic mass is 36.9659 amu and the percentage abundance is 24.47%. Calculate the atomic weight of chlorine.

Solution:

The contribution of each isotope to the relative atomic weight is determined using the "weighted average" method. Each of the isotopic masses is multiplied by the fractional abundance and then the products summed.

For ^{35}Cl: (0.7553) x (34.9689 amu) = 26.41201 amu
(calculator answer)
= 26.41 amu
(correct answer)

For ^{37}Cl: (0.2447) x (36.9659 amu) = 9.0455557 amu
(calculator answer)
= 9.046 amu
(correct answer)

Remember significant figures. Significant figures must be taken into account in any calculation involving experimentally determined numbers.

Summing the products gives:

At. Wt. = 26.41 amu + 9.046 amu = 35.456 amu
(calculator answer)
= 35.46 amu
(correct answer)

Since 26.41 is known only to the hundredths place, the answer can be expressed only to the hundredths place.

8. <u>Understand and state the periodic law</u> (section 5.8).

A statement of the periodic law is: "When all elements are arranged in order of increasing atomic numbers, elements with similar properties occur at periodic (regularly recurring) intervals."

The periodic law concept enables chemists to systematize

the study of the properties of the elements.

9. <u>Understand the rationale behind the organization of the periodic table. Relate the terms period and group to the periodic table. List the general information given in the periodic table</u> (section 5.8).

The periodic table is a graphical representation of the periodic behavior described by the periodic law. All elements with similar chemical properties are placed underneath each other, forming a column in the table.

<u>Periods</u> in the periodic table are horizontal rows of elements. The periods are numbered sequentially starting at the top of the table.

<u>Groups</u> in the periodic table are vertical columns of elements. All of the elements in a group, by design, have similar chemical properties. Roman numerals and the letters A and B are used in designating groups.

Each "box" in the periodic table corresponds to a different element. The following items of information are found within the box: (1) elemental symbol, (2) atomic number and (3) atomic weight. Note that mass numbers are not one of the pieces of information found within a box.

10. <u>Write a correct symbol for an ion (including charge) given the number of protons, neutrons and electrons present</u> (section 5.9).

Under certain circumstances an atom may gain or lose electrons. Loss of negative electrons results in a positively charged atom, since it now has more positive protons than negative electrons. Gain of negative electrons results in a negatively charged atom, since it now has more negative electrons than positive protons.

The charged atoms formed by the loss or gain of electrons are called ions. An <u>ion</u> is an atom (or group of atoms) that is electrically charged as a result of an excess or deficiency of electrons.

The charge on an ion is directly related to the number of electrons gained or lost. Loss of one, two or three electrons gives, respectively, ions with a +1, +2 and +3 charge. Gain of one, two or three electrons gives, respectively, ions with a -1, -2 and -3 charge.

The notation for indicating the charge on an ion is a

superscript placed to the right of the symbol of the element. Some examples of ion symbols are:

positive ions -- K^+, Ca^{2+}, Al^{3+}

negative ions -- Br^-, S^{2-}, P^{3-}

Notice that only a plus or minus is used to indicate a charge of one, instead of using 1+ or 1-. Also notice that in multiply-charged ions the number preceeds the charge sign; that is, the correct notation for a charge of plus two is 2+ rather than +2.

Example:

Give the symbol for each of the following ions:

a) the ion formed when a magnesium atom loses 2 electrons
b) the ion formed when a sulfur atom gains 2 electrons

Solution:

a) A neutral magnesium atom contains 12 protons and 12 electrons. A magnesium ion would still contain 12 protons, but would have only 10 electrons since 2 electrons were lost.

12 protons   = 12 + charges
10 electrons = 10 - charges
Net charge   = 2+

The symbol for the ion is Mg^{2+}.

b) The atomic number of sulfur is 16. Thus, there are 16 protons and 16 electrons in a neutral sulfur atom. The gain of 2 electrons increases the number of electrons to 18.

16 protons   = 16 + charges
18 electrons = 18 - charges
Net charge   = 2-

The symbol for the ion is S^{2-}.

PROBLEM SET

1. Identify each of the following as a heteroatomic or homoatomic molecule.

   a) HBr

   b) N_2

   c) H_2S

   d) CCl_4

   e) S_8

   f) I_2

   g) CH_3Cl

   h) H_2

2. How many atoms of each kind are represented by the following formulas:

   a) $(NH_4)_3PO_4$

   b) CO_2

   c) SO_3

   d) $Mg(C_2O_4)_2$

   e) $C_6H_{12}O_6$

   f) CH_5N

   g) $Al_2(SO_3)_3$

   h) P_2O_5

3. Complete the following table:

Particle	Electrical charge	Relative mass	Location within atom
Neutron			
Proton			
Electron		1	

4. Determine the number of neutrons, protons and electrons contained in an atom of the following isotopes:

   a) $^{15}_{7}N$

   b) $^{18}_{8}O$

   c) $^{34}_{16}S$

   d) $^{63}_{29}Cu$

   e) $^{137}_{56}Ba$

   f) $^{55}_{25}Mn$

   g) $^{127}_{53}I$

   h) $^{209}_{83}Bi$

5. Write the complete symbols ($^{A}_{Z}$symbol) for each of the atoms:

   a) the hydrogen isotope containing 2 neutrons
   b) the lead isotope containing 125 neutrons

66   The Atom and its Structure

c) the bromine isotope containing 45 neutrons
d) the mercury isotope containing 121 neutrons

6. Nitrogen occurs in nature in two isotopic forms, $^{14}_{7}N$ and $^{15}_{7}N$. The isotopic mass of $^{14}_{7}N$ is 14.0031 amu and its percentage abundance is 99.63%. For $^{15}_{7}N$ the isotopic mass is 15.0001 amu and the percentage abundance is 0.37. Calculate the atomic weight of nitrogen.

7. Uranium occurs in nature in three isotopic forms, $^{234}_{92}U$, $^{235}_{92}U$ and $^{238}_{92}U$. The isotopic masses and percentage abundances for the three isotopes are 234.0409, 235.0439, 238.0508, and 0.0057, 0.72, 99.27, respectively. Calculate the atomic weight of uranium.

8. Identify, with the help of the periodic table, each of the following elements:

   a) located in period 3 and group IIA
   b) located in group IVA and period 5
   c) located in group VIIA and period 2
   d) located in period 4 and group IIIB

9. Give the symbol for each of the following ions:

   a) the ion formed when a silver atom loses one electron
   b) the ion formed when an atom of barium loses two electrons
   c) the ion formed when an atom of aluminum loses three electrons
   d) the ion formed when an atom of fluorine gains one electron
   e) the ion formed when an atom of oxygen gains two electrons
   f) the ion formed when an atom of phosphorus gains three electrons

10. Calculate the number of protons and electrons in each of the following ions:

   a) $_{17}Cl^{-}$

   b) $_{4}Be^{2+}$

   c) $_{19}K^{+}$

   d) $_{50}Sn^{4+}$

MULTIPLE CHOICE EXERCISES

1. Which of the following is not a postulate of atomic theory?

    a) chemical change involves union, separation or rearrangement of atoms
    b) different "types" of atoms exist
    c) atoms change size during chemical change
    d) atoms are indestructible during chemical change

2. Which of the following is not a postulate of atomic theory?

    a) all matter is made up of atoms
    b) all atoms of a given "type" are identical to each other
    c) only whole atoms can participate in chemical reactions
    d) atoms of a given "type" differ significantly from atoms of other "types"

3. The smallest "piece" of a compound that retains the properties of the compound is:

    a) an atom
    b) an element
    c) a molecule
    d) a nucleus

4. Molecules of compounds differ from molecules of elements in that only compound molecules:

    a) are homoatomic
    b) are heteroatomic
    c) may be heteroatomic or homoatomic
    d) contain atoms

5. Which of the following is a homoatomic molecule?

    a) HI
    b) P_4
    c) HCl
    d) CO_2

6. Which of the following formulas contains 6 atoms of oxygen?

   a) Na_2SO_4
   b) $Al(NO_3)_3$
   c) $Ba(NO_3)_2$
   d) $Al_2(SO_4)_3$

7. Consider the formulas $COCl_2$ and $CoCl_2$. It is true that they:

   a) contain the same number of elements
   b) are identical in meaning
   c) contain different numbers of atoms
   d) are both examples of homoatomic molecules

8. Which of the following statements concerning a proton is not true?

   a) possesses a positive charge
   b) is the heaviest of the fundamental subatomic particles
   c) is found in the nucleus
   d) can be called a nucleon

9. All atoms are neutral because

   a) all atoms contain neutrons
   b) the charge of the protons balances the charge of the electrons
   c) the nucleus is neutral
   d) the number of nucleons equals the number of electrons

10. The nucleus of an atom

    a) accounts for almost all of the weight of an atom
    b) contains only neutrons
    c) is negatively charged
    d) contains both protons and electrons

11. An atom with a mass number of 33 and an atomic number of 16 would be correctly represented by which of the following symbols:

    a) $^{33}_{16}As$ c) $^{33}_{16}S$
    b) $^{16}_{33}As$ d) $^{16}_{33}S$

12. An isotope has an atomic number of 20 and a mass number of 45. This means that the nucleus of the atom contains:

   a) 45 protons and 20 neutrons
   b) 20 protons and 45 neutrons
   c) 20 protons and 25 neutrons
   d) 25 protons and 20 neutrons

13. Which of the following statements is correct for $^{30}_{14}Si$?

   a) contains more protons than neutrons
   b) contains more electrons than protons
   c) contains an equal number of protons and neutrons
   d) contains more neutrons than electrons

14. Isotopes must:

   a) have the same mass number
   b) contain the same number of neutrons
   c) have the same atomic number
   d) have a different number of electrons

15. In general, the atomic weight of an element is:

   a) the weight of any single atom of that element
   b) the weight of a particular isotope of that element
   c) the average weight of atoms of that element compared to a standard
   d) the average weight of a particular isotope of that element compared to a standard

16. An element is made up of two naturally occurring isotopes. 20% of the atoms of the element have a relative weight of 20.0 amu and the other 80% a relative weight of 22.0 amu. What is the atomic weight of the element?

   a) 20.0 amu
   b) 20.4 amu
   c) 21.6 amu
   d) 42.0 amu

17. The periodic law states that when elements are arranged in order of _____ their properties repeat themselves at regular intervals.

   a) increasing atomic number
   b) decreasing atomic number
   c) increasing atomic weight
   d) decreasing atomic weight

18. In the periodic table, elements with similar chemical properties are found together

    a) in vertical columns
    b) in horizontal rows
    c) in zig-zag arrangements
    d) in sequential groups of four

19. Which one of the following elements is in both period 6 and group IIIA?

    a) $_{15}P$
    b) $_{33}As$
    c) $_{49}In$
    d) $_{81}Tl$

20. Which of the following pieces of information about an element is not listed on a periodic table?

    a) atomic number
    b) mass number
    c) atomic weight
    d) chemical symbol

21. Formation of a positive ion is the result of an atom:

    a) losing one or more protons
    b) losing one or more electrons
    c) gaining one or more protons
    d) gaining one or more electrons

22. The number of electrons in a $_{56}^{137}Ba^{2+}$ ion is:

    a) 54
    b) 56
    c) 58
    d) 135

23. The complete symbol for an ion containing 8 protons, 9 neutrons and 10 electrons would be:

    a) $_{8}^{17}O^{2+}$
    b) $_{8}^{17}O^{2-}$
    c) $_{10}^{19}Ne^{2-}$
    d) $_{10}^{19}Ne^{2+}$

ANSWERS TO PROBLEMS SET

1. a) heteroatomic
    b) homoatomic
    c) heteroatomic
    d) heteroatomic
    e) homoatomic
    f) homoatomic
    g) heteroatomic
    h) homoatomic

2. a) 3 nitrogen, 12 hydrogen, 1 phosphorus, 4 oxygen
    b) 1 carbon, 2 oxygen
    c) 1 sulfur, 3 oxygen
    d) 1 magnesium, 4 carbon, 8 oxygen
    e) 6 carbon, 12 hydrogen, 6 oxygen
    f) 1 carbon, 5 hydrogen, 1 nitrogen
    g) 2 aluminum, 3 sulfur, 9 oxygen
    h) 2 phosphorus, 5 oxygen

3.
neutron	0	1839	nucleus
proton	+1	1836	nucleus
electron	−1	1	extra nuclear region

4. a) 8 neutrons, 7 protons, 7 electrons
    b) 10 neutrons, 8 protons, 8 electrons
    c) 18 neutrons, 16 protons, 16 electrons
    d) 34 neutrons, 29 protons, 29 electrons
    e) 81 neutrons, 56 protons, 56 electrons
    f) 30 neutrons, 25 protons, 25 electrons
    g) 74 neutrons, 53 protons, 53 electrons
    h) 126 neutrons, 83 protons, 83 electrons

5. a) $^{3}_{1}H$ c) $^{80}_{35}Br$

    b) $^{207}_{82}Pb$ d) $^{201}_{80}Hg$

6. $^{14}_{7}N$: 0.9963 x 14.0031 amu = 13.951289 amu
                              (calculator answer)
                         = 13.95 amu (correct answer)

    $^{15}_{7}N$: 0.0037 x 15.0001 amu = 0.05550037 amu
                              (calculator answer)
                         = 0.056 amu (correct answer)

    At. Wt. = (13.95 + 0.056) amu = 14.006 amu
                                  (calculator answer)
                             = 14.01 amu
                                (correct answer)

7. $^{234}_{92}U$: 0.000057 x 234.0409 amu = 0.01334033 amu
                                    (calculator answer)
                               = 0.013 amu
                                (correct answer)

$^{235}_{92}$U: 0.0072 x 235.0439 amu = 1.6923161 amu
                                      (calculator answer)
                               = 1.7 amu (correct answer)

$^{238}_{92}$U: 0.9927 x 238.0508 amu = 236.31303 amu
                                      (calculator answer)
                               = 236.3 amu (correct answer)

At. Wt. = (0.013 + 1.7 x 236.3) amu = 238.013 amu
                                      (calculator answer)
                               = 238.0 amu
                                 (correct answer)

8. a) $_{12}Mg$ c) $_9F$

    b) $_{50}Sn$ d) $_{21}Sc$

9. a) Ag^+ d) F^-

    b) Ba^{2+} e) O^{2-}

    c) Al^{3+} f) P^{3-}

10. a) 17 protons and 18 electrons (−1 charge indicates gain
       of an electron
    b) 4 protons and 2 electrons (+2 charge indicates loss of
       2 electrons)
    c) 19 protons and 18 electrons (+1 charge indicates loss
       of 1 electron)
    d) 50 protons and 46 electrons (+4 charge indicates loss
       of 4 electrons)

ANSWERS TO MULTIPLE CHOICE ANSWERS

1. c   8. b

2. b   9. b

3. c   10. a

4. b   11. c

5. b   12. c

6. c   13. d

7. c   14. c

15. c

16. c

17. a

18. a

19. d

20. b

21. b

22. a

23. b

CHAPTER 6
Compounds: Their Formulas and Names

REVIEW OF CHAPTER OBJECTIVES

1. <u>State the law of definite proportions, and show how composition data for compounds is treated to verify this law</u> (section 6.1).

    The law of definite proportions states that in a pure compound the elements are always present in the same definite proportion by mass.

    For example, any sample of pure copper oxide, no matter how it is analyzed, will always show the same ratio of copper to oxygen.

    The following example shows how to treat experimental data to verify the law of definite proportions.

    Three samples of water (H_2O) of different masses and from different sources, are decomposed to yield hydrogen and oxygen. The results of the decompositions are as follows:

	Sample Mass	Mass of Hydrogen Produced	Mass of Oxygen Produced
Sample 1	2.000	0.224	1.776
Sample 2	5.634	0.632	5.002
Sample 3	8.542	0.955	7.587

    Solution:

    By calculating the percentage oxygen contained in each sample, the law of definite proportions can be verified. All the oxygen percentages should come out the same within experimental error if the law is obeyed.

Percentage oxygen = $\dfrac{\text{mass of oxygen produced}}{\text{mass of sample}}$ x 100

Sample 1: %O = $\dfrac{1.776}{2.000}$ x 100 = 88.8% (calculator answer)

= 88.80% (correct answer to
4 significant figures)

Sample 2: %O = $\dfrac{5.002}{5.634}$ x 100 = 88.782393%
(calculator answer)
= 88.78% (correct answer)

Sample 3: %O = $\dfrac{7.587}{8.542}$ x 100 = 88.819948%
(calculator answer)
= 88.82% (correct answer)

The three percentages are the same to three significant figures, thus proving the law of definite proportions. There is variation in the fourth significant figure due to measuring uncertainty and rounding in the original experimental data.

2. <u>Distinguish between ionic and molecular compounds;</u> <u>between binary and ternary compounds</u> (section 6.3).

Ionic compounds have structures based upon infinite 3-dimensional arrangements of positive and negative ions.

Most ionic compounds contain both a metallic and a nonmetallic element. It is not necessary to memorize which elements are metals and which are nonmetals, since this information can be obtained from a periodic table. Those elements to the left of the step-like line are metals, and those to the right are nonmetals (see Figure 6-1 on page 118 of the textbook). (Notice that the location of hydrogen presents an exception; it is a nonmetal.)

Most molecular compounds contain only nonmetallic elements.

Example: Classify each of the following compounds as molecular or ionic:

a) KBr   b) CO_2 c) $NaNO_3$ d) H_2S

Solution:

Using the periodic table, determine if each of the elements present in the compound is a metal or a nonmetal. If one is a metal and the other is a

nonmetal, the compound is ionic. If both elements
are nonmetallic, the compound is molecular.

a) Potassium (K) is a metal. Bromine (Br) is a
nonmetal. Therefore, the compound is ionic.

b) Carbon (C) is a nonmetal. Oxygen (O) is a
nonmetal. Therefore, the compound is molecular.

c) Sodium (Na) is a metal. Both nitrogen (N) and
oxygen (O) are nonmetals. The compound contains
a metal and 2 nonmetals and is ionic. If the
compound contains at least one metal and one
nonmetal, it will be ionic. But, it may contain
more than one of either.

d) Hydrogen (H) is a nonmetal. Remember the
location of hydrogen is an exception. Sulfur (S)
is a nonmetal. Therefore, the compound is
molecular.

A second classification scheme for compounds is based on
the number of elements present in a compound.

Binary compounds contain just two different elements.
Any number of atoms of the two elements may be present,
but only two elements may be present.

The compounds HCl, H_2O, NH_3, CH_4 are all binary compounds
since they all contain only two different elements. The
compound CH_4, even though it contains one atom of carbon
(C) and 4 atoms of hydrogen (H), is still a binary
compound because it contains only the two different
elements, carbon and hydrogen.

Ternary compounds contain three or more different
elements. The compounds $Al(NO_3)_3$, H_3PO_4 and NaOH are all
ternary compounds.

3. <u>Given the name or symbol, state the charge(s) on common
monoatomic ions</u> (section 6.4).

To have success in the study of chemistry it is necessary
to learn the charge on the most common monoatomic ions,
since knowing charge magnitude is a prerequisite for
writing formulas of ionic compounds and for naming the
compounds.

To help systematize the process of learning ionic charges
for monoatomic ions, they are classified into three
categories:

Compounds: Their Formulas and Names 77

1. Ionic of nonmetallic elements

    A. Nonmetals in group VIIA of the periodic table are always 1-.

    B. Nonmetals in group VIA of the periodic table are always 2-.

    C. Nonmetals in group VA of the periodic table are always 3-.

    These generalizations for nonmetals are true only for binary ionic compounds.

2. Ions of metallic elements where the metal forms only one type of ion

    A. Metals in group IA of the periodic table are always 1+.

    B. Metals in group IIA of the periodic table are always 2+.

    C. Aluminum (Al) is always a 3+.

    D. Silver (Ag) is always a 1+.

    E. Zinc (Zn) and cadmium (Cd) are always 2+.

    These rules hold true for all compounds containing these elements.

3. Ions of metallic elements where the metal forms more than one type of ion.

    It is not possible to predict the magnitude of the charge on the ions of these metals from their position in the periodic table. The ionic charges for these common variable charge metals must just be learned.

Element	Ions Formed
chromium	Cr^{2+} and Cr^{3+}
cobalt	Co^{2+} and Co^{3+}
copper	Cu^{+} and Cu^{2+}
gold	Au^{+} and Au^{3+}
iron	Fe^{2+} and Fe^{3+}
lead	Pb^{2+} and Pb^{4+}
manganese	Mn^{2+} and Mn^{3+}
tin	Sn^{2+} and Sn^{4+}

4. <u>Write formulas for binary ionic compounds given the combining elements</u> (section 6.5).

The number of electrons lost by a metal must always equal the number of electrons gained by the nonmetal when an ionic compound is formed. Consequently, ionic compounds are always neutral. That is, they have no net charge.

The procedures for writing formulas for ionic compounds are based on this required total charge balance between positive and negative ions. The ratio in which positive and negative ions combine is that ratio which achieves charge neutrality for the resulting compound.

The following steps can be used as a systematic approach to writing formulas of compounds:

Step 1: Write the symbols of the metal and nonmetal side by side in the order metal/nonmetal.

Step 2: Determine the charge on the metal ion and the nonmetal ion, using the previously discussed rules.

Step 3: The third step is to determine the lowest common multiple of the two charges. The signs on the two numbers are ignored in determining the lowest common multiple. The lowest common multiple is the smallest number into which the numbers of the charges of the two elements in a compound can be divided to give a whole number.

Step 4: Determine the subscripts, if any, used in the formula. The subscript of the metal is determined by dividing its charge number into the lowest common multiple; the subscript of the

nonmetal is determined by dividing its charge
number into the lowest common multiple.

Step 5:: Write the formula using the subscripts
determined in step 4.

Step 6: Check to see that the total charge on the
compound is zero. This is done by multiplying
the subscript of the metal times the charge on
the metal ion. Multiply the subscript of the
nonmetal times the charge on the nonmetal ion
and obtain the algebraic sum. If the formula
has been written correctly the net charge will
be zero.

Example: Determine the formula for the resulting ionic
compound when each of the following pairs of
elements (a metal and a nonmetal) interact:

a) K and N
b) Al and O
c) Sn and S (Sn is present as Sn^{4+})

Solution:

a) **Step 1**: KN

Step 2: K is in group IA so it is 1+.
N is in group VA so it is 3-.

Step 3: The lowest common multiple is (3 x 1)
= 3.

Step 4: The subscript for K is 3.

$$\frac{\text{lowest common multiple}}{\text{charge number of K}} = \frac{3}{1} = 3$$

The subscript for N is 1.

$$\frac{\text{lowest common multiple}}{\text{charge number of N}} = \frac{3}{3} = 1$$

Step 5: The correct formula is K_3N.

Step 6: Total metal charge   = (3 x 1+) = 3+
Total nonmetal charge = (1 x 3-) = 3-
                    net charge   = 0

Therefore, formula is written correctly.

b) **Step 1:** AlO

Step 2: Al is always a 3+.
O is in group VIA so it is 2-.

Step 3: Lowest common multiple is (3 x 2) = 6.
Six is the smallest number which 2 and 3
can be divided by and give a whole
number.

Step 4: Subscript for Al $\dfrac{6}{3} = 2$

Subscript for O $\dfrac{6}{2} = 3$

Step 5: The correct formula is Al_2O_3.

Step 6: Total metal charge $= (2 \times 3+) = 6+$
Total nonmetal charge $= (3 \times 2-) = \underline{6-}$
net charge $= 0$

Therefore, formula is correctly written.

c) **Step 1:** SnS

Step 2: We are told the charge on the Sn is 4+.
For the metals with variable charges,
you must be told the charge unless you
are given the name of the compound. You
will learn to do this later in the
chapter. S is in VIA so it is a 2-.

Step 3: The lowest common multiple is 4 since it
is the smallest number which 2 and 4
will go into. If we used the previous
method of determining the lowest common
multiple (2 x 4) = 8, our subscripts
would be twice as large as they should
be. Subscripts should always be the
smallest possible numbers. You may use
this method to determine lowest common
multiples, but be sure to check that it
is in fact the lowest common multiple;
and, if not, reduce to the lowest common
multiple.

Step 4: Subscript for Sn $\dfrac{4}{4} = 1$

$$\text{Subscript for S} \quad \frac{4}{2} = 2$$

Step 5: The correct formula is SnS_2.

Step 6:
$$\begin{aligned}
\text{Total metal charge} &= (1 \times 4+) = 4+ \\
\text{Total nonmetal charge} &= (2 \times 2-) = 4- \\
\text{net charge} &= 0
\end{aligned}$$

Therefore, formula is correctly written.

5. <u>Given the formula of a binary ionic compound, write its name and vice versa</u> (section 6.6).

For purposes of naming binary ionic compounds, classification is in one of two categories:

1. Binary ionic compounds containing a fixed-charge metal.

2. Binary ionic compounds containing a variable-charge metal.

Binary ionic compounds containing a fixed-charge metal are named using the following procedure:

1. Give the full name of the metallic element first.

2. Then, as a separate word, give the stem of the nonmetallic element name and the suffix -ide.

An important fact to remember is that all binary compound names will end in "ide." If you remember this fact, when you see a name ending in "ide," you will immediately recognize it as being a binary compound. And, if you are naming a compound containing only two elements, you will know that to be named correctly it must end in "ide." The stem name of the nonmetal is always the first few letters of the nonmetal's name. That is, the name of the nonmetal with its ending chopped off.

You should commit the following names and formulas for the most common nonmetallic ions to memory.

Name of Ion	Formula
bromide	Br^-
carbide	C^{4-}
chloride	Cl^-
fluoride	F^-
hydride	H^-
iodide	I^-
nitride	N^{3-}
oxide	O^{2-}
phosphide	P^{3-}
sulfide	S^{2-}

Example: Name the following binary ionic compounds, all of which contain a fixed-charge metal.

a) CaO   b) $AlCl_3$ c) Na_3P

Solution:

a) The metal is calcium and the nonmetal is oxygen. Oxygen with its -ide ending is an oxide. The name of the compound is calcium oxide.

b) The metal is aluminum. The nonmetal is chlorine. Chlorine with its -ide ending is chloride. The name of the compound is aluminum chloride.

Note that no mention is made in the name of the subscript 3 found after the symbol for chlorine. The name of an ionic compound never contains any reference to the subscript numbers contained in the formula. Since aluminum is a fixed-charge metal, there is only one ratio in which fluorine and aluminum may combine. Thus, just telling the names of the elements in the compound is adequate nomenclature.

c) The metal is sodium and the nonmetal is phosphorus. Phosphorus, with its -ide ending, is phosphide. The correct name is sodium phosphide.

An important fact which many students fail to appreciate is that chemical nomenclature is very specific. There should be no possibility for misinterpretation. There can be only one correct name for a formula. If you write

Compounds: Their Formulas and Names   83

a name for which someone else could write a different formula, it is not correct.

Binary ionic compounds containing a variable-charge metal are named using the following procedure:

1. Give the full name of the metallic element first. Follow immediately with a Roman numeral enclosed in parentheses to indicate the magnitude of the charge on the metal ion. This Roman numeral is considered to be part of the metal ion's name. For example, iron (II) ion.

2. Then, as a separate word, give the name of the nonmetal with its -ide ending.

The following example illustrates the naming of binary ionic compounds containing a variable-charge metal ion.

Example: Name the following binary ionic compounds, all of which contain a variable-charge metal ion:

   a) SnO b) SnO_2 c) Fe_2S_3

Solution:

   a) The charge on the variable-charge metal ion must be determined before the name can be written. This is done by using the charge on the nonmetal ion which is always known and the fact that all ionic compounds are neutral. Oxygen is in VIA so it has a -2 charge. Therefore, for the compound to be neutral the charge on the tin ion must be 2+.

$$2+ \ 2- \ = 0$$
$$Sn \ \ O$$

   The name of the compound is tin(II) oxide.

   b) Oxygen is in group VIA so it has a -2 charge. The total charge due to the oxide ion is determined by multiplying the subscript (2) by the charge on a single oxide ion (2-) (2 x 2- = 4-). Therefore, for the compound to be neutral the charge on the tin ion in this compound must be 4+.

$$4+ \ 2(2-) \ = 0$$
$$Sn \ \ \ O_2$$

The name of the compound is tin(IV) oxide.

Comparing the results of a) and b), you should see the importance of correct nomenclature. If either compound had been incorrectly named as tin oxide, a person writing the formula from just the incorrect name would not know whether to write SnO of SnO_2.

c) Sulfur is in group VIA so it is a 2–. The total charge due to the 3 sulfide ions in the compound is $3 \times 2 = 6-$. Therefore, for the compound to be neutral the total charge due to the 2 iron ions in the compound is 6+.

$$6+ \quad 3(2-) = 0$$
$$Fe_2 \quad S_3$$

When the name of a compound is written, the Roman numeral indicates the magnitude of the charge on a single metal ion. In this compound, there are 2 iron ions with a total charge of 6+, so each iron ion would have a charge of 3+.

$$\frac{\text{total metal ion charge}}{\text{metal ion subscript}} = \frac{6+}{2} = 3+$$

The name of the compound is iron(III) sulfide.

Remember that the Roman numeral gives the magnitude of the charge on a single iron ion in the compound. Many students forget this meaning and think the Roman numeral indicates the subscript of the metal ion in the formula of the compound. As a result, they name it incorrectly as iron(II) oxide which would have a formula of FeO. Also notice there is no reference to the sign of the charge. Metals are always positive so it is unnecessary and incorrect to name it as iron(III+) oxide. It is equally incorrect to name it as iron(3+) oxide or iron(3) oxide, since the magnitude of the charge on the ion is correctly indicated using a Roman numeral.

The procedure for writing a formula from the name is the same six steps outlined in objective 4. The following example will illustrate the use.

Example: Write the formula for the following binary ionic compounds:

a) barium iodide
b) cobalt(II) fluoride
c) gold(III) oxide

Solution:

a) **Step 1:** BaI

   Step 2: Barium is a fixed-charge metal found in group IIA so its charge is 2+. Iodine is a nonmetal found in group VIIA so the charge on the iodide ion is 1-.

   Step 3: The lowest common multiple is (2 x 1) = 2.

   Step 4: The subscript for Ba is 1. $\dfrac{2}{2} = 1$

   The subscript for I is 2. $\dfrac{2}{1} = 2$

   Step 5: The correct formula is BaI_2.

   Step 6:
   Total metal charge   = (1 x 2+) = 2+
   Total nonmetal charge = (2 x 1-) = 2-
   net charge = 0

b) **Step 1:** CoF

   Step 2: Cobalt is a variable-charge metal whose charge is given by the Roman numeral (II) so its charge is 2+. Fluorine is a nonmetal found in group VIIA so the charge on a fluoride ion is 1-.

   Step 3: The lowest common multiple is (2 x 1) = 2.

   Step 4: The subscript for Co is 1. $\dfrac{2}{2} = 1$

   The subscript for F is 2. $\dfrac{2}{1} = 2$

   Step 5: The correct formula is CoF_2.

   Step 6:
   Total metal charge   = (1 x 2+) = 2+
   Total nonmetal charge = (2 x 1-) = 2-
   net charge = 0

c) **Step 1:**   AuO

    Step 2:   The charge on gold is 3+ given by the Roman numeral in the name. Oxygen is in group VIA so the charge on the oxide ion is 2-.

    Step 3:   The lowest common multiple is (3 x 2) = 6.

    Step 4:   The subscript for Au is 2. $\frac{6}{3} = 2$

              The subscript for O is 3. $\frac{6}{2} = 3$

    Step 5:   The correct formula is Au_2O_3.

    Step 6:   Total metal charge $= (2 \times 3^+) = 6+$
                Total nonmetal charge $= (3 \times 2-) = \underline{6-}$
                       net charge $=$ 0

6.   <u>Recognize the names and symbols of common polyatomic ions</u> (section 6.7).

A polyatomic ion is a group of atoms tightly bound together which have acquired a charge. Polyatomic ions are very stable species, generally maintaining their identity during chemical reactions. Numerous ionic compounds exist in which the positive or negative ion (sometimes both) is polyatomic.

Polyatomic ions are not molecules and do not occur alone. Instead, they are always found associated with other ions of opposite charge. Polyatomic ions are pieces of compounds, not compounds. Polyatomic ions are always charged species.

The inability to recognize the presence of polyatomic ions (both by name and formula) in a compound is a major stumbling block for many chemistry students. Effort must be put forth to avoid this obstacle. There is no easy way for learning the common polyatomic ions. Memorization is required.

Table 6-5 on page 128 in the textbook gives the formulas and names of the most commonly encountered polyatomic ions. Notice that two of the polyatomic ions in Table 6-5, OH^- (hydroxide) and CN^- (cyanide), have an "ide" ending. These names represent exceptions to the rule that the suffix "ide" is used only in naming binary compounds.

7. <u>Write formulas for ionic compounds containing polyatomic ions given the identity of the combining ions</u> (section 6.8).

Formulas for ionic compounds containing polyatomic ions are determined using the same six steps used for binary ionic compounds containing monoatomic ions.

Two conventions not encountered previously in formula writing often arise when writing polyatomic-ion-containing formulas. They are:

1. When more than one polyatomic ion of a given kind is required in a formula, the polyatomic ion is enclosed in parentheses and a subscript placed outside the parentheses to indicate the number of polyatomic ions needed to achieve neutrality of the compound.

2. To preserve the identity of polyatomic ions, the same elemental symbol may be used more than once in a formula.

The following example contains illustrations of both these new conventions.

Example: Determine the formulas for the ternary ionic compounds containing the following pairs of ions:

   a) Na^+ and CN^-
   b) Mg^{2+} and PO_4^{3-}
   c) NH_4^+ and NO_3^-

Solution:

   a) **Step 1:** Na^+ CN^-

   In this example, the cyanide polyatomic ion is written in place of a monoatomic nonmetallic ion. However, the convention of positive first followed by negative is still followed.

   Step 2: Na^+ is a 1+ and CN^- is a 1-.

   Step 3:   The lowest common multiple is 1.

   Step 4:   The subscript for Na^+ is 1, and the subscript for CN^- is 1.

Step 5: The name of the compound is sodium cyanide. This compound ends in "ide," even though it is not a binary compound, since the name of the CN^- ion is cyanide. This is one of the exceptions to the "ide" ending rule.

Step 6:
$$\begin{aligned}
\text{Total metal ion charge} &= 1 \times 1+ = 1+ \\
\text{Total polyatomic ion charge} &= 1 \times 1- = \underline{1-} \\
\text{net charge} &= 0
\end{aligned}$$

b) **Step 1:** $Mg^{2+} \; PO_4^{3-}$

Step 2: Mg is 2+ and PO_4 is 3-.

Step 3: The least common multiple is $(2 \times 3) = 6$.

Step 4: The subscript for Mg^{2+} is 3. $\qquad \dfrac{6}{2} = 3$

The subscript for PO_4^{3-} is 2. $\qquad \dfrac{6}{3} = 2$

Step 5: The correct formula is $Mg_3(PO_4)_2$.

The parentheses are necessary since 2 PO_4^{3-} ions are needed.

Step 6:
$$\begin{aligned}
\text{Total metal ion charge} &= (3 \times 2+) = 6+ \\
\text{Total polyatomic ion charge} &= (2 \times 3-) = \underline{6-} \\
\text{net charge} &= 0
\end{aligned}$$

c) **Step 1:** $NH_4^+ \; NO_3^-$

This is an example where both the positive ion (NH_4^+) and the negative ion (NO_3^-) are polyatomic. Notice that the convention of positive ion followed by negative ion is still followed.

Step 2: NH_4 is a 1+ and NO_3 is a 1-.

Step 3: The lowest common multiple is $(1 \times 1) = 1$.

Step 4: The correct formula is NH_4NO_3. Notice that the NH_4^+ and NO_3^- still retain their separate identities in the correct formula. If all the nitrogen were combined and the formula written as $N_2H_4O_3$ it would not be possible to tell that the compound contained an ammonium ion (NH_4^+) and a nitrate ion (NO_3^-). It is also incorrect to write the formula as $(NH_4)(NO_3)$. When only one polyatomic ion is needed, no parentheses are necessary.

8. <u>Given the formula of an ionic compound containing polyatomic ions, write its name and vice versa</u> (section 6.9).

The names of ionic compounds containing polyatomic ions are derived in a manner similar to that for binary ionic compounds.

1. Name the positive ion first, polyatomic or metallic (including a Roman numeral indicating the charge of the variable-charge metallic ion, when needed).

2. Then, as a separate word, give the name of the negative ion (polyatomic or monoatomic nonmetal with its "ide" ending).

The following example illustrates the rules.

Example: Name the following compounds, which contain one or more polyatomic ions:

   a) Na_2SO_4 b) $Cu(NO_3)_2$ c) $FePO_4$

Solution:

   a) The metal is sodium. The polyatomic ion is sulfate. The correct name is sodium sulfate.

   Sodium is found in group IA and therefore has a fixed charge of 1+, so the name sodium(I) sulfate would be incorrect. Remember Roman numerals are used only for variable-charge metals. The name sodium(II) sulfate is incorrect because the Roman numeral is being used to refer to the number of sodium ions contained in the compound. Roman numerals are used only to refer to the charge on a variable-charge metal ion. No reference is ever made to the number of ions contained in an ionic compound.

   The importance of memorizing and recognizing polyatomic ions should also be apparent. If you do not recognize an SO_4 as the polyatomic sulfate ion you could not correctly name this compound. Many students incorrectly name this as sodium sulfide when they do not recognize the sulfate ion (SO_4^{2-}). If they remembered that the "ide" ending is reserved for binary compounds, they would realize this could not be correct. The formula for the binary compound sodium sulfide is Na_2S. Or, all else failing, they name it sodium sulfur oxide, which is entirely incorrect because there is no such polyatomic ion as a sulfur oxide. And it contains 3 elements, Na, S and O, so it is not binary and could not end in "ide."

b) The metal is copper. Copper must be recognized as a variable-charge metal. The difficulty arises in determining what its charge is.

The polyatomic ion NO_3 (nitrate ion) with a charge of 1- must be recognized from memorization. Using this information and the fact that there are 2 nitrate ions it is determined that the total negative charge is 2^-.

$$2(1-)$$
$$Cu(NO_3)_2$$

Therefore, for the compound to be neutral, the charge on the copper ion must be 2+.

$$2+ \ 2(1-) = 0$$
$$Cu \ (NO_3)_2$$

The correct name of the compound is copper(II) nitrate.

The name copper nitrate is incorrect since it does not indicate the charge of the variable-charge copper and could mean $Cu(NO_3)_2$ or $CuNO_3$. Copper(I) nitrate is incorrect because the copper is 2+ not 1+ as the Roman numeral (I) would indicate.

c) The metal ion is iron. Iron is a variable-charge metal whose charge is obtained from the polyatomic phosphate ion (PO_4^{3-}).

$$3-$$
$$FePO_4$$

If the phosphate ion is a 3-, then the iron must

be 3+ in order for the compound to remain neutral.

$$3+ \quad 3- = 0$$
$$Fe \quad PO_4$$

The correct name is iron(III) phosphate.

If named incorrectly as iron phosphate it would not be known whether the compound's formula was $Fe_3(PO_4)_2$, iron(II) phosphate, or $FePO_4$, iron (III) phosphate.

The procedure for writing the formula from the name of a compound containing a polyatomic ion uses the same six steps we have used in previous examples.

The following example illustrates their use for polyatomic-ion-containing compounds.

Example: Write the formula for the following compounds which contain one or more polyatomic ions.

    a) ammonium acetate
    b) cobalt(III) sulfate
    c) potassium hydroxide

Solution:

a) **Step 1:** $NH_4 \quad C_2H_3O_2$

    The importance of memorizing the polyatomic ions should immediately be apparent. You must recognize from the name that the ammonium ion is (NH_4^+) and the acetate ion is ($C_2H_3O_2^-$).

Step 2:   The charge on the ammonium ion must be recalled as a 1+. The charge on the acetate ion must be recalled as a 1-.

Step 3:   The least common multiple is (1 x 1) = 1.

Step 4:   The subscript for the ammonium ion is 1. The subscript for the acetate ion is 1.

Step 5:   The correct formula is $NH_4C_2H_3O_2$.

    The formulas $(NH_4)C_2H_3O_2$, $NH_4(C_2H_3O_2)$ and $(NH_4)(C_2H_3O_2)$ are all incorrect since no parentheses are needed. Parentheses are only used when more than one polyatomic ion is required to achieve neutrality which is not the case in this example.

Notice that each polyatomic ion retains its individual identity which would not be true if the formula were written incorrectly as $NH_7C_2O_4$.

Step 6: Total positive charge = (1 x 1+) = $\underline{1+}$
Total negative charge = (1 x 1-) = $\underline{1-}$
net charge = 0

b) Step 1: $Co\ SO_4$

Step 2: The positive ion is the variable-charge metal ion whose charge is 3+ given by the Roman numeral (III). It must be recalled that the sulfate ion is SO_4^{2-} with a charge of 2-.

The importance of memorization cannot be over-emphasized. If you have not memorized that a sulfate ion is (SO_4^{2-}), you will not recognize it and could just as easily put SO_3^{2-} (sulfite ion) or S^{2-} (sulfide ion).

Step 3: The lowest common multiple is 3 x 2 = 6.

Step 4: The subscript for cobalt is 2. $\frac{6}{3} = 2$

The subscript for sulfate is 3. $\frac{6}{2} = 3$

Step 5: The correct formula is $Co_2(SO_4)_3$.

The SO_4^{2-} must be enclosed in parentheses since more than one is needed. The Co is not enclosed in parentheses since monoatomic ions are never enclosed in parentheses.

Step 6: Total positive charge = (2 x 3+) = $\underline{6+}$
Total negative charge = (3 x 2-) = $\underline{6-}$
net charge = 0

Compounds: Their Formulas and Names 93

c) **Step 1:**   K OH

   Step 2:   The charge on the potassium ion is 1+.
   The charge on the hydroxide ion is 1-.

   Step 3:   The lowest common multiple is $(1 \times 1) = 1$.

   Step 4:   The subscript for K is 1. The subscript
   for OH is 1.

   Step 5:   The formula is KOH. The OH is not
   enclosed in parentheses since only one
   is needed.

   Step 6:   Total positive charge = $(1 \times 1+)$ = 1+
   Total negative charge = $(1 \times 1-)$ = 1-
   net charge = 0

9. Given the formula of a binary molecular compound write
    its name and vice versa (section 6.11).

The names of binary molecular compounds will consist of
two words with the words having the following general
formats:

First word: (prefix) + (full name of the first element)

Second word: (prefix) + (stem of name of second
element) + (ide)

The prefixes are Greek numerical prefixes indicating the
number of atoms of each element present in a molecule of
a compound. The Greek numerical prefixes for the numbers
one through ten are given in Table 6-6 on page 132 of
your textbook.

Example: Name the following binary molecular compounds:

a) CO_2 b) S_2O c) PCl_3

Solution:

a) The elements present are carbon and oxygen.
Where there is only one atom of the first element
present, it is a common practice to omit the
beginning prefix mono- for that element.
Following this guideline, we have the name carbon
dioxide.

b) There are 2 atoms of sulfur and one atom of
oxygen per molecule. Thus, the name is disulfur

monoxide.

   c) The prefix for three is tri-. Thus, this
      compound has the name phosphorus trichloride.

Writing formulas for binary molecular compounds given
their names is a very easy task. The Greek prefixes in
the names of such compounds tell you exactly how many
atoms of each kind are present. For example, the
compound diphosphorus trioxide has the formula P_2O_3 (di-
and tri-).

10. <u>Recognize acids from other molecular compounds,
    distinguish between monoxyacids and oxyacids, and name
    both types of acids given the formula or vice versa</u>
    (section 6.12).

Many molecular hydrogen-containing compounds dissolve in
water to give solutions with properties which are
distinctly different from the pure compound before it was
dissolved in water. These solutions are called acids.

Not all hydrogen-containing molecular compounds produce
acids when dissolved in water. Those that do can be
recognized by their formulas which have hydrogen written
as the first element in the formula.

HCl, H_2S, H_2SO_4 and HNO_3 all function as acids when
dissolved in water.

NH_3, CH_4, PH_3 and H_2O are examples of compounds which do
not form acids. Note that the H is not written first in
these formulas except for H_2O (water) which is an
exception.

Because of the difference in properties of acids and the
pure molecular compounds from which acids are produced,
acids are given different names. For naming purposes,
acids are classified into two categories:

1. Nonoxyacids -- water solutions of molecular compounds
    composed of hydrogen and one or more nonmetals other
    than oxygen.

2. Oxyacids -- water solutions of molecular compounds
    made up of hydrogen, some other element and oxygen.

Nonoxyacids are named by modifying the name of the pure
molecular compound as follows:

1. The hydrogen is completely dropped, and the prefix hydro- added to the stem of the name of the nonmetal.

2. The suffix -ide on the stem of the name of the nonmetal is replaced with the suffix -ic.

3. The word acid is added to the end of the name (as a separate word).

The following example illustrates the use of these rules.

Example: Name the following binary molecular compounds as acids:

    a) hydrogen chloride (HCl)
    b) hydrogen sulfide (H_2S)
    c) hydrogen telluride (H_2Te)

Solutions:

  a) 1. Drop the hydrogen and add the prefix hydro-, hydrochloride.

     2. Replace the suffix -ide with -ic, hydrochloric.

     3. Add the word acid as a separate word. Hydrochloric acid is the correct name for a water solution of HCl.

  b) 1. Hydrosulfide

     2. Hydrosulfuric

     For sulfur-containing acids the "ur" from sulfur is reinserted in the acid name for ease of pronunciation.

     3. Hydrosulfuric acid

  c) 1. Hydrotelluride

     2. Hydrotelluric

     3. Hydrotelluric acid

Oxyacids are ternary molecular compounds all having two elements, hydrogen and oxygen, in common.

Oxyacid molecules, when dissolved in water, break apart

to give ions. The positive ions produced are H^+ ions, the ion characteristic of all acids. The negative ions produced are the same polyatomic ions we have dealt with earlier.

Formulas for common oxyacids are H_2SO_4, HNO_3, H_3PO_4 and H_2CO_3.

Names for oxyacids are derived from the names of the polyatomic ions produced when the acid molecules break into ions in solution. Polyatomic ion names are modified in the following manner to give oxyacid names:

1. When the polyatomic ion produced in solution has a name ending in -ate, the -ate ending is dropped, the suffix -ic is added in its place and the word acid is added.

2. When the polyatomic ion produced in solution has a name ending in -ite, the -ite ending is dropped, the suffix -ous is added in its place and the word acid is added.

The following example illustrates the use of these rules.

Example: Name the following ternary compounds as acids:

a) H_2SO_4 b) $HClO_4$ c) H_2SO_3

Solution:

a) The polyatomic ion present in solutions of this acid is the sulfate ion (SO_4^{2-}). Removing the -ate ending from the word sulfate, replacing it with the suffix -ic and adding the word acid gives the name sulfuric acid for this oxyacid.

Note the close similarity of the two acid names, hydrosulfuric acid and sulfuric acid. Many students have difficulty distinguishing between the two. If you thoroughly understand acid naming, they should present no difficulty. The prefix hydro- in hydrosulfuric acid indicates the non-oxyacid (H_2S), while the absence of a prefix in sulfuric acid indicates the oxyacid (H_2SO_4). The prefix hydro- is reserved for and used only with nonoxyacids.

b) The polyatomic ion present in solutions of this acid is the perchlorate ion (ClO_4^-). Removing the -ate ending from the word perchlorate, replacing it with the suffix -ic, and adding the word acid gives the name perchloric acid.

Note that if a polyatomic ion's name contains a prefix, it is carried over to the acid name since it is a definite part of the polyatomic ion's name.

c) The polyatomic ion produced when this compound dissolves in water is sulfite (SO_3^{2-}). Note that two closely related (and often confused) polyatomic ions exist -- sulfite (SO_3^{2-}) and sulfate (SO_4^{2-}). We are dealing here with sulfite. For polyatomic ions whose names end in -ite, the -ite suffix is replaced by an -ous suffix and the word acid added. This compound is thus named sulfurous acid.

Again from this example, you should realize the importance of memorizing and recognizing the polyatomic ions.

You should also be able to write the formula for an acid given its name as illustrated in the following example.

Example: Give the formulas for the following acids:

a) hydrobromic acid
b) phosphoric acid
c) hypochlorous acid

Solution:

a) The hydro- prefix indicates that this is a nonoxyacid. In forming the acid name, the -ide ending of the nonmetal ion was changed to -ic. Applying the reverse process, we obtain the bromide ion which is (Br^-). The formula is HBr since it is an acid which indicates that the positive ion present is H^+. Acids must be neutral just as all compounds, which is true here. Hydrogen ions are 1+ and bromide ions are 1-, and there is one of each present in the correct formula HBr.

b) The -ic ending of the acid name without a hydro- prefix indicates that the formula will contain a polyatomic ion whose name ends in -ate. The phosphorus-containing polyatomic ion whose name ends in -ate is phosphate (PO_4^{3-}). The hydrogen present for formula writing purposes is H^+ ion. Combining these ions (H^+) and (PO_4^{3-}) in the right ratio to give a neutral compound gives the formula H_3PO_4 for phosphoric acid.

c) The -ous ending of the acid name indicates that the formula will contain a polyatomic ion whose name ends in -ite. (Note: Many students confuse the two prefixes, hydro- and hypo- when dealing with this compound. The prefixes do not have the same meaning. Hydro- is used for nonoxyacids, which is not the case here. Hypo- is part of the name for the polyatomic ion whose name is hypochlorite ion.) This ion has the formula ClO^-. Combining the H^+ and ClO^- ions in the correct ratio to give a neutral compound results in the formula $HClO$ for hypochlorous acid.

PROBLEMS SET

1. Two different samples of a pure compound (containing elements X and Y) were analyzed with the following results:

   Sample I: 35.47 grams of compound yielded 20.51 grams of A and 15.00 grams of B.

   Sample II: 44.38 grams of compound yielded 25.66 grams of A and 18.72 grams of B.

   Show that this data is consistent with the law of definite proportions.

2. Identify the following compounds as ionic or molecular and classify each as a binary or ternary compound.

   a) KCl e) NH_4Cl

   b) CCl_4 f) CH_2Cl_2

   c) N_2O_4 g) $Ba(CN)_2$

   d) Na_2S h) HCl

3. Give the charges on the monoatomic ions which the following elements will form:

a) nitrogen
b) iron
c) sulfur
d) iodine

e) copper
f) P
g) K
h) Mg

4. Write formulas for the compounds formed from the following ions:

a) Li^+ and S^{2-}
b) Ba^{2+} and Br^-
c) Al^{3+} and Cl^-
d) Cd^{2+} and N^{3-}

e) Ag^+ and I^-
f) Sn^{2+} and S^{2-}
g) Co^{2+} and P^{3-}
h) Mg^{2+} and O^{2-}

5. Write a correct formula for the binary ionic compound formed by combining:

a) sodium and chlorine
b) beryllium and oxygen
c) gold (the 3+ ion) and oxygen
d) aluminum and nitrogen
e) iron (the +2 ion) and iodine
f) lithium and bromine
g) barium and sulfur
h) zinc and fluorine

6. Name the following binary ionic compounds:

a) $MgBr_2$
b) Na_2O
c) CdO
d) KCl

e) CoO
f) Cu_2O
g) $SnCl_2$
h) PbO_2

7. Write formulas for the following binary ionic compounds:

a) lithium chloride
b) silver sulfide
c) barium bromide
d) magnesium nitride

e) aluminum phosphide
f) iron(III) fluoride
g) zinc iodide
h) manganese(III) sulfide

8. Give names for the following polyatomic ions:

a) NO_2^-

b) SO_3^{2-}

c) $C_2H_3O_2^-$

d) ClO_3^-

e) CrO_4^{2-}

f) NH_4^+

g) HSO_4^-

h) PO_3^{3-}

9. Write formulas (including charge) for the following polyatomic ions:

a) nitrate ion
b) sulfate ion
c) hydroxide ion
d) cyanide ion

e) carbonate ion
f) phosphate ion
g) hydrogen carbonate ion
h) permanganate ion

10. Write formulas for the compounds formed between these positive and negative ions:

a) K^+ and ClO_4^-

b) NH_4^+ and Cl^-

c) Ca^{2+} and SO_4^{2-}

d) Cu^{2+} and OH^-

e) Pb^{4+} and PO_4^{3-}

f) Ag^+ and $Cr_2O_7^{2-}$

g) Au^+ and $C_2H_3O_2^-$

h) Fe^{3+} and $C_2O_4^{2-}$

11. Write formulas for the ionic compound formed by combining:

a) barium ion and nitrate ion
b) iron(III) ion and hydroxide ion
c) aluminum ion and carbonate ion
d) sodium ion and perchlorate ion
e) potassium ion and phosphate ion
f) ammonium ion and sulfide ion
g) cobalt(II) ion and dihydrogen phosphate ion
h) ammonium ion and acetate ion

12. Name the following compounds:

a) $Co_2(SO_4)_3$

b) NH_4CN

c) $KHCO_3$

d) Na_3PO_4

e) $Pb(OH)_2$

f) Ag_2CO_3

g) $Au(NO_3)_3$

h) $AlPO_3$

Compounds: Their Formulas and Names   **101**

13. Write formulas for the following compounds:

   a) copper(II) sulfate
   b) ammonium nitrate
   c) barium nitrite
   d) chromium(II) hydrogen phosphate
   e) zinc acetate
   f) sodium nitrite
   g) sodium dichromate
   h) potassium permanganate

14. Name the following binary molecular compounds:

   a) SF_6 e) ICl
   b) OF_2 f) N_2O
   c) SO_2 g) CF_4
   d) P_4O_6 h) NH_3

15. Which of the following can be classified as an acid?

   a) $H_2Cr_2O_7$ e) HCN
   b) CH_4 f) NH_3
   c) $H_2C_2O_4$ g) H_2O
   d) $HClO$ h) AsH_3

16. Classify the following as an oxyacid or a nonoxyacid:

   a) HF e) H_2Se
   b) $HClO_4$ f) HNO_2
   c) HCN g) H_3PO_4
   d) H_2SO_3 h) HCl

17. Name the following compounds as acids:

   a) $HC_2H_3O_2$ e) HI
   b) H_3PO_3 f) H_2S
   c) H_2CO_3 g) $H_2C_2O_4$
   d) $HClO_4$ h) H_2CrO_4

18. Write formulas for the following acids:

   a) hydrobromic acid   e) phosphoric acid
   b) chlorous acid   f) hydrocyanic acid
   c) nitric acid   g) hydrochloric acid
   d) nitrous acid   h) hydroselenic acid

MULTIPLE CHOICE EXERCISES

1. According to the law of definite proportions, if a sample of a compound, A, contains 10 grams of sulfur and 5 grams of oxygen, then another sample of A which contains 20 grams of sulfur must contain:

    a) 10 g of oxygen
    b) 5 g of oxygen
    c) 2.5 g of oxygen
    d) 1.0 g of oxygen

2. Which of the following is an ionic compound?

    a) $BaCl_2$

    b) HCl

    c) CCl_4

    d) NH_3

3. Which of the following is a ternary compound?

    a) $MgCl_2$

    b) LiCN

    c) NaCl

    d) $AlCl_3$

4. Nonmetals in groups VIA and VIIA of the periodic table would, respectively, be expected to form ions with charges of:

    a) +6 and +7
    b) −6 and −7
    c) −1 and −2
    d) −2 and −1

5. The correct formula for the ionic compound containing Al^{3+} and S^{2-} ions would be:

    a) AlS

    b) Al_3S_2

    c) Al_2S_3

    d) AlS_2

6. The Roman numeral (III) in the name iron(III) oxide indicates that:

   a) there are three times as many iron atoms present as oxygen atoms
   b) there are three iron atoms per molecule
   c) there are three oxygen atoms per molecule
   d) iron atoms present are ions with a +3 charge

7. Which of the following compounds is named without using a Roman numeral?

   a) $AgCl$

   b) $CuCl_2$

   c) $CoCl_3$

   d) $TiCl_4$

8. Which of the following names is paired with an incorrect formula?

   a) sodium chloride -- $NaCl$

   b) calcium oxide -- CaO

   c) lithium nitride -- Li_3N

   d) magnesium sulfide -- MgS_2

9. Which of the following formulas is paired with an incorrect name?

   a) $AuBr_3$ -- gold(III) bromide

   b) Cu_2O -- copper(II) oxide

   c) Fe_2O_3 -- iron(III) oxide

   d) $CoCl_2$ -- cobalt(II) chloride

10. Which of the following compounds contains 3 atoms per formula unit?

   a) magnesium nitride
   b) aluminum oxide
   c) calcium bromide
   d) sodium fluoride

11. Which of the following statements about polyatomic ions is correct?

    a) all have names which end in -ate
    b) all must contain oxygen
    c) negative ions are more common than positive ions
    d) all carry a charge of two

12. Which of the following polyatomic ions contains the greatest number of oxygen atoms?

    a) ammonium
    b) sulfate
    c) nitrate
    d) chlorate

13. Which of the following polyatomic ions has a negative two charge associated with it?

    a) cyanide
    b) carbonate
    c) phosphate
    d) hydroxide

14. The correct formula for the compound formed by combining iron(II) ions and phosphate ions is:

    a) Fe_2PO_4
    b) $Fe_3(PO_4)_2$
    c) $(Fe)_2(PO_4)_3$
    d) $FePO_4$

15. Which of the following names is paired with a correct formula?

    a) sodium chlorate -- Na_2CO_3
    b) ammonium nitrate -- NH_4CN
    c) lithium phosphate -- Li_2PO_4
    d) aluminum carbonate -- $Al_2(CO_3)_3$

16. Which of the following compounds has a name which does not contain the prefix di-?

   a) S_2O
   b) NO_2
   c) OF_2
   d) Li_2O

17. Which of the following acids would have a name containing the suffix -ous?

   a) HNO_3
   b) H_2SO_3
   c) H_3PO_4
   d) H_2CO_3

18. Which of the following is not an oxyacid?

   a) sulfuric acid
   b) hypochlorous acid
   c) hydrofluoric acid
   d) nitrous acid

19. Which of the following acids is correctly named?

   a) H_2SO_4 -- hydrosulfuric acid
   b) H_3PO_3 -- phosphoric acid
   c) $HClO$ -- hydrochlorous acid
   d) HNO_2 -- nitrous acid

20. Which of the following compounds contains the greatest number of atoms per formula unit?

   a) aluminum nitrate
   b) sodium cyanide
   c) hydrobromic acid
   d) tetraphosphorus tetrasulfide

ANSWERS TO PROBLEMS SET

1. Sample I: %A = $\dfrac{20.51}{35.47}$ x 100 = 57.823513 (calculator answer)

$$= 57.82 \text{ (correct answer)}$$

Sample II: $\%A = \dfrac{25.66}{44.38} \times 100 = 57.818837$ (calculator answer)

$$= 57.82 \text{ (correct answer)}$$

Since the percentage A in sample I equals percentage A in sample II, the law of definite proportions is followed.

2. a) ionic, binary   e) ionic, ternary
    b) molecular, binary   f) molecular, ternary
    c) molecular, binary   g) ionic, ternary
    d) ionic, binary   h) molecular, binary

3. a) 3-   e) 1+ or 2+

    b) 2+ or 3+   f) 3-

    c) 2-   g) 1+

    d) 1-   h) 2+

4. a) Li_2S e) AgI

    b) $BaBr_2$ f) SnS

    c) $AlCl_3$ g) Co_3P_2

    d) Cd_3N_2 h) MgO

5. a) $NaCl$ e) FeI_2

    b) BeO f) $LiBr$

    c) Au_2O_3 g) BaS

    d) AlN h) ZnF_2

6. a) magnesium bromide   e) cobalt(II) oxide
    b) sodium oxide   f) copper(I) oxide
    c) cadmium oxide   g) tin(II) chloride
    d) potassium chloride   h) lead(IV) oxide

7. a) $LiCl$ e) AlP

    b) Ag_2S f) FeF_3

    c) $BaBr_2$ g) ZnI_2

    d) Mg_3N_2 h) Mn_2S_3

8. a) nitrite ion   e) chromate ion
    b) sulfite ion   f) ammonium ion
    c) acetate ion   g) hydrogen sulfate ion or bisulfate ion
    d) chlorate ion   h) phosphite ion

9. a) NO_3^-
   b) SO_4^{2-}
   c) OH^-
   d) CN^-
   e) CO_3^{2-}
   f) PO_4^{3-}
   g) HCO_3^-
   h) MnO_4^-

10. a) $KClO_4$
    b) NH_4Cl
    c) $CaSO_4$
    d) $Cu(OH)_2$
    e) $Pb_3(PO_4)_4$
    f) $Ag_2Cr_2O_7$
    g) $AuC_2H_3O_2$
    h) $Fe_2(CrO_4)_3$

11. a) $Ba(NO_3)_2$
    b) $Fe(OH)_3$
    c) $Al_2(CO_3)_3$
    d) $NaClO_4$
    e) K_3PO_4
    f) $(NH_4)_2S$
    g) $Co(H_2PO_4)_2$
    h) $NH_4C_2H_3O_2$

12. a) cobalt(III) sulfate
    b) ammonium cyanide
    c) potassium bicarbonate or potassium hydrogen carbonate
    d) sodium phosphate
    e) lead(II) hydroxide
    f) silver carbonate
    g) gold(III) nitrate
    h) aluminum phosphite

13. a) $CuSO_4$
    b) NH_4NO_3
    c) $Ba(NO_2)_2$
    d) $CrHPO_4$
    e) $Zn(C_2H_3O_2)_2$
    f) $NaNO_2$
    g) $Na_2Cr_2O_7$
    h) $KMnO_4$

14. a) sulfur hexafluoride
    b) oxygen difluoride
    c) sulfur dioxide
    d) tetraphosphorus hexaoxide
    e) iodine monochloride
    f) dinitrogen monoxide
    g) carbon tetrafluoride
    h) ammonia

15. a, c, d, e

16. Nonoxyacid -- a, c, e, h
    Oxyacid -- b, d, f, g

17. a) acetic acid
    b) phosphorus acid
    c) carbonic acid
    d) perchloric acid
    e) hydroiodic acid
    f) hydrosulfuric acid
    g) oxalic acid
    h) chromic acid

18. a) HBr   e) H_3PO_4
    b) $HClO_2$ f) HCN
    c) HNO_3 g) HCl
    d) HNO_2 h) H_2Se

ANSWERS TO MULTIPLE CHOICE EXERCISES

1. a   11. c

2. a   12. b

3. b   13. b

4. d   14. b

5. c   15. d

6. d   16. d

7. a   17. b

8. d   18. c

9. b   19. d

10. c   20. a

CHAPTER 7
Chemical Calculations I:
The Mole Concept
and Chemical Formulas

REVIEW OF CHAPTER OBJECTIVES

1. Calculate the formula weight of a substance given its formula and a table of atomic weights (section 7.1).

    Determining the formula weight of a substance is an important first step in many chemical calculations. The formula weight is the sum of the atomic weights of all of the atoms in the formula unit. It is necessary to count the number of atoms of each element that are found in the formula unit, and then add up their total weight.

    The formula weight of NO, which has one atom of nitrogen and one atom of oxygen, is:

    ```
    1 N = 1 x 14.01 amu  = 14.01 amu
    1 O = 1 x 16.00 amu  = 16.00 amu
    formula weight of NO = 30.01 amu
    ```

    The formula weight of NO_2, which has one nitrogen atom and two oxygen atoms, is:

    ```
    1 N:  1 x 14.01 amu   = 14.01 amu
    2 O:  2 x 16.00 amu   = 32.00 amu
    formula weight of NO₂ = 46.01 amu
    ```

    The formula weight of oxygen gas, O_2, which has two oxygen atoms per molecule, is:

    ```
    2 O:  2 x 16.00 amu = 32.00 amu
    ```

2. Calculate the percentage composition of any component in a compound, from the formula of the compound and a table of atomic weights (section 7.2).

    Percentage means parts per hundred, and percentage

composition means the percentage by weight of an element in a compound. To determine the percentage of oxygen in the compound N_2O_3, one first determines the formula weight of the compound. This was found to be 76.02 amu (on the previous page).

The general formula used for percentage composition is:

% element = $\dfrac{\text{weight of element in one formula unit}}{\text{formula weight}}$ x 100

So,

% oxygen = $\dfrac{48.00 \text{ amu}}{76.02 \text{ amu}}$ x 100 = 63.141278%
                                         (calculator answer)
                                       = 63.14% (correct answer)

3. <u>Interpret the mole as a counting unit and calculate the number of particles in a given number of moles of a substance</u> (section 7.3).

<u>The mole</u> is the chemist's counting unit. It is 6.02 x 10^{23} particles. The number 6.02 x 10^{23} is also called Avogadro's number.

1 mole of objects = 6.02 x 10^{23} objects
                             = Avogadro's number of objects

The mole will be used often in many different chemical problems in this book; it is important to understand the difference between an atom, a molecule and a mole.

An atom is the smallest unit of an element. A molecule contains two or more atoms chemically bound together. Individual atoms or molecules are much too small to see or weigh. However, a mole of atoms (6.02 x 10^{23} atoms) can be seen and weighed. Similarly, a mole of molecules (6.02 x 10^{23} molecules) can also be seen and weighed.

The definition of a mole, 1 mole of a substance = 6.02 x 10^{23} units of that substance, can be used to generate conversion factors for use in problem solving situations. The two conversion factors are:

$$\dfrac{1 \text{ mole}}{6.02 \text{ x } 10^{23} \text{ objects}} \quad \text{and} \quad \dfrac{6.02 \text{ x } 10^{23} \text{ objects}}{1 \text{ mole}}$$

Note how these conversion factors are used in the following examples.

Example: How many atoms are in 1.23 moles of Zn?

Solution: 1.23 ~~moles Zn~~ x $\left(\dfrac{6.02 \times 10^{23} \text{ atoms Zn}}{1 \text{ ~~mole Zn~~}}\right)$

$= 7.4046 \times 10^{23}$ atoms Zn (calculator answer)

$= 7.40 \times 10^{23}$ atoms Zn (correct answer)

Example: How many molecules are in 0.257 moles of CH_4?

Solution: 0.257 ~~moles CH~~$_4$ x $\left(\dfrac{6.02 \times 10^{23} \text{ molecules } CH_4}{1 \text{ ~~mole CH~~}_4}\right)$

$= 1.54714 \times 10^{23}$ molecules CH_4 (calculator answer)

$= 1.55 \times 10^{23}$ molecules CH_4 (correct answer)

4. <u>Interpret the mole as a variable mass unit and calculate the mass of a given number of moles of a substance</u> (section 7.4)

How much a mole of a substance weighs depends on the identity of the substance. Just as a dozen watermelon, a dozen grapes and a dozen helium atoms will have different masses, so also will a mole of watermelons, a mole of grapes and a mole of helium atoms have different masses. Thus, the mass of a mole, or the <u>molar mass</u>, is not one set number; it varies, being different for each different substance.

The molar mass of an <u>element</u>, when the element is in <u>atomic form</u>, is equal to a mass in grams numerically equal to the <u>atomic weight</u> of the element.

One mole of He weighs 4.00 g (at. wt. = 4.00 amu)
One mole of Li weighs 6.94 g (at. wt. = 6.94 amu)
one mole of Au weighs 196.97 g (at. wt. = 196.97 amu)

The molecular form of an element will have a different molar mass than will its atomic form. When an <u>element</u> is in a <u>molecular form</u> its molar mass is a mass in grams numerically equal to its <u>molecular weight</u>.

One mole of O_2 weighs 32.00 g (mol. wt. = 32.00 amu)

One mole of H_2 weighs 2.00 g (mol. wt. = 2.00 amu)

Note that these molar masses are double what they would be if the elements were in atomic form since the molecular forms involve diatomic molecules.

The molar mass of a <u>compound</u> is a mass in grams that is numerically equal to the <u>formula weight</u> of the compound.

One mole NO_2 weighs 46.01 g (form. wt. = 46.01 amu)

One mole N_2O_3 weighs 76.02 g (form. wt. = 76.02 amu)

One mole NaCl weighs 58.45 g (form. wt. = 58.45 amu)

The molar mass of a substance may be used as a conversion factor to convert moles to grams or grams to moles.

Example: 185 moles of NO_2 weighs how many grams?

Solution: The molar mass of NO_2 is 46.0 grams. From this statement, two conversion factors may be written:

$$\frac{46.0 \text{ grams } NO_2}{1 \text{ mole } NO_2} \text{ and } \frac{1 \text{ mole } NO_2}{46.0 \text{ grams } NO_2}$$

The first of these conversion factors is used in solving the problem.

$$185 \text{ moles } NO_2 \text{ x } \frac{46.0 \text{ grams } NO_2}{1 \text{ mole } NO_2} =$$

$$= 8510 \text{ grams } NO_2 \text{ (calculator and correct answer)}$$

Example: 185 grams of NO_2 contains how many moles of NO_2?

Solution: The conversion factor for this problem is the reciprocal of the one used previously.

$$185 \text{ grams } NO_2 \text{ x } \frac{1 \text{ mole } NO_2}{46.0 \text{ grams } NO_2}$$

$$= 4.0217391 \text{ moles } NO_2 \text{ (calculator answer)}$$

$$= 4.02 \text{ moles } NO_2 \text{ (correct answer)}$$

5. <u>Recognize the mass ratio between substances of which equal numbers of particles are present</u> (section 7.5).

Samples of two or more pure substances (element or compound) which have mass ratios equal to the ratios of their atomic or formula weights will contain the same number of particles (atoms, molecules, or formula units). Thus, to obtain samples of elements or compounds containing a like number of particles, we merely weigh out quantities (in any units) whose mass ratio is numerically equal to the ratio of the substance's atomic or molecular weights.

Example: Do 2.00 grams of Au and 1.00 gram of Ag contain equal numbers of atoms?

Solution: From the periodic table, the ratio of the atomic weights of the two elements is:

$$\frac{\text{atomic weight Au}}{\text{atomic weight Ag}} = \frac{197 \text{ amu}}{108 \text{ amu}}$$

$$= 1.8240741 \text{ (calculator answer)}$$
$$= 1.82 \text{ (correct answer)}$$

The ratio of the masses of the two given samples is:

$$\frac{\text{mass Au}}{\text{mass Ag}} = \frac{2.00 \text{ grams}}{1.00 \text{ grams}} = 2 \text{ (calculator answer)}$$
$$= 2.00 \text{ (correct answer)}$$

Since the ratio of the atomic weights and the ratio of the given masses are not the same, equal numbers of atoms are not present.

Example: Do 4.73 grams of CO_2 and 3.01 grams of CO contain equal numbers of molecules?

Solution: If the mass ratio and the molecular weight ratios are the same (to three significant figures), then these amounts of substances contain equal numbers of molecules.

Step 1:   To three significant figures, the molecular weight of CO_2 is 44.0 amu, and the molecular weight of CO is 28.0 amu.

$$\frac{\text{molecular weight } CO_2}{\text{molecular weight CO}} = \frac{44.0 \text{ amu}}{28.0 \text{ amu}}$$

$$= 1.5714286 \text{ (calculator answer)}$$
$$= 1.57 \text{ (correct answer)}$$

Step 2:   The ratio of the masses of the two given samples is:

$$\frac{\text{mass } CO_2}{\text{mass CO}} = \frac{4.73 \text{ g}}{3.01 \text{ g}} = 1.5714286 \text{ (calculator answer)}$$
$$= 1.57 \text{ (correct answer)}$$

Since the molecular weight ratio and the given
mass ratio are the same, equal numbers of
molecules are present.

6. <u>Interpret the subscripts in a chemical formula in terms
of the number of moles of the various elements in one
mole of the substance</u> (section 7.6).

The numerical subscripts in a formula tell the number of
atoms of each element which are present in one formula
unit of the substance.

The formula H_2SO_4 states that the smallest unit of this
compound contains:

> two atoms of hydrogen
> one atom of sulfur, and
> four atoms of oxygen

The subscripts in a formula also tell the number of moles
of atoms of the various elements which are present in one
mole of the substance.

One <u>mole</u> of H_2SO_4 contains:

> two moles of hydrogen atoms
> one mole of sulfur atoms, and
> four moles of oxygen atoms

7. <u>Given the formula of a substance, calculate any one of
the following from another of the quantities</u> (section
7.7):

<u>a) mass of substance, b) moles of substance, c) number
of particles of substance, d) atoms of any element,
e) mass of any element, f) moles of any element.</u>

The three quantities most often calculated in chemical
problems are <u>grams</u> of a substance, <u>moles</u> of a substance
or number of <u>particles</u> of a substance (atoms, molecules
or formula units).

<u>Avogadro's number</u> provides a relationship between the
number of particles of a substance and the number of
moles of the same substance.

<u>Molar mass</u> provides a relationship between the number of
grams of a substance and the number of moles of the same
substance.

<u>Molar interpretation of chemical formula subscripts</u>

provides a relationship between the number of moles of a substance and the number of moles of its component parts.

The first two of the above concepts are sufficient to work problems where information is given about a particular substance (moles, particles or grams) and additional information is asked for concerning this <u>same substance</u>. The following diagram, which also appears on page 160 in the textbook (Figure 7-1), arranges these concepts in a manner that is very useful in determining the sequence of conversion factors needed to solve problems:

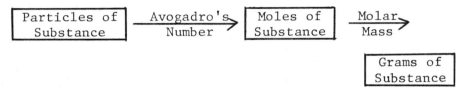

All three of the previously listed concepts are needed for calculations where information (moles, particles or grams) is given about a particular substance and information is asked for concerning a <u>related substance</u>

In working problems of this type the following diagram (also found as Figure 7-2 on page 160 of the textbook) is very useful:

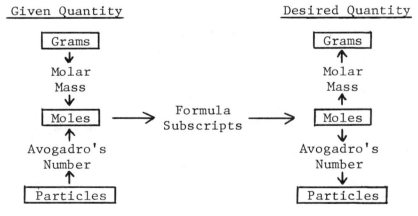

Both Figures 7-1 and 7-2 are "road maps". From them the "pathway" (sequence of conversion factors needed) for solving a given problem is easily determined. The arrows between boxes give the concept to be used when effecting the indicated unit changes (grams to moles, moles to particles, etc.).

Example: How many molecules are present in 5.00 grams of H_2O?

Solution:

The given quantity is 5.00 g of H_2O and the desired quantity is molecules (particles) of H_2O.

$$5.00 \text{ g } H_2O = ? \text{ molecules } H_2O$$

In "box language" this is a grams-to-particles problem. Using Figure 7-1 we find that the correct pathway for the problem is:

$$\boxed{\begin{array}{c}\text{Grams}\\ H_2O\end{array}} \xrightarrow[\text{Mass}]{\text{Molar}} \boxed{\begin{array}{c}\text{Moles}\\ H_2O\end{array}} \xrightarrow[\text{Number}]{\text{Avogadro's}} \boxed{\begin{array}{c}\text{Molecules}\\ H_2O\end{array}}$$

The molar mass of H_2O is 18.0 g. Avogadro's number has the value 6.02×10^{23}. Both of these numbers are needed in working the problem.

Translating this pathway into conversion factors gives

$$5.00 \text{ g } H_2O \times \left(\frac{1 \text{ mole } H_2O}{18.0 \text{ g } H_2O}\right) \times \left(\frac{6.02 \times 10^{23} \text{ molecules } H_2O}{1 \text{ mole } H_2O}\right)$$

$$= 1.6722222 \times 10^{23} \text{ molecules } H_2O \text{ (calculator answer)}$$

$$= 1.67 \times 10^{23} \text{ molecules } H_2O \text{ (correct answer)}$$

Example: Find the mass, in grams, of one atom of sulfur.

Solution:

This problem is just the opposite of the previous example -- a particles-to-grams problem rather than a grams-to-particles problem. The pathway is thus just the reverse of that previously used.

$$\boxed{\text{Particles S}} \xrightarrow[\text{Number}]{\text{Avogadro's}} \boxed{\text{Moles S}} \xrightarrow[\text{Mass}]{\text{Molar}} \boxed{\text{Grams S}}$$

The molar mass of S is 32.1 g (obtained from the atomic weight of S). The set-up for the problem, using dimensional analysis, is

$$1 \text{ atom S} \times \left(\frac{1 \text{ mole S}}{6.02 \times 10^{23} \text{ atoms S}}\right) \times \left(\frac{32.1 \text{ g S}}{1 \text{ mole S}}\right) =$$

$$= 5.3322259 \times 10^{-23} \text{ g S (calculator answer)}$$
$$= 5.33 \times 10^{-23} \text{ g S (correct answer)}$$

The answer is given to three significant figures because the molar mass and Avogadro's number are used to three significant figures. The one atom of sulfur is an exact (counted) number, and does not limit the number of significant figures in the answer.

Example: How many silver atoms are in 1.50 g Ag_3PO_4?

Solution:

This problem is fundamentally different from the two previous examples. In the previous problems information was given about a substance and additional information was asked for concerning the same substance. In this problem information is given about one substance (Ag_3PO_4) and new information is desired about a related substance (Ag). For this "related" type of problem Figure 7-2, rather than Figure 7-1, is used as the "road map".

The problem is a grams-to-particles problem. The appropriate pathway for solving it is:

The set-up, using dimensional analysis, is:

$$1.50 \text{ g } Ag_3PO_4 \times \left(\frac{1 \text{ mole } Ag_3PO_4}{429 \text{ g } Ag_3PO_4} \right) \times \left(\frac{3 \text{ moles Ag atoms}}{1 \text{ mole } Ag_3PO_4} \right)$$

$$\times \left(\frac{6.02 \times 10^{23} \text{ Ag atoms}}{1 \text{ mole Ag atoms}} \right) = 6.314685 \times 10^{21} \text{ Ag atoms}$$
$$\text{(calculator answer)}$$
$$= 6.31 \times 10^{21} \text{ Ag atoms}$$
$$\text{(correct answer)}$$

Note how the subscript 3 in the formula Ag_3PO_4 was used in obtaining the 2nd conversion factor. One mole of Ag_3PO_4 contains 3 moles of Ag atoms.

8. Calculate the empirical formula of a substance from percentage composition information (section 7.8).

The empirical (or simplest) formula of a compound gives the smallest whole-number ratio of the numbers of atoms of each element that are present in the formula unit.

To determine the empirical formula, the percentage composition (or the weights of each element found in a given weight) of the compound must be changed to moles of elements. Then, the smallest whole-number ratio of moles of elements is determined. These small whole numbers are the subscripts in the empirical formula.

Example: Find the empirical formula of a compound which is 70.0% iron and 30.0% oxygen.

Solution:

100.0 grams of this compound contains 70.0 g Fe and 30.0 g O.

First we determine the number of moles of each element present:

$$70.0 \text{ g Fe} \times \left(\frac{1 \text{ mole Fe}}{55.8 \text{ g Fe}}\right) = 1.2544803 \text{ moles Fe}$$
(calculator answer)
$$= 1.25 \text{ moles Fe}$$
(correct answer)

$$30.0 \text{ g O} \times \left(\frac{1 \text{ mole O}}{16.0 \text{ g O}}\right) = 1.875 \text{ moles O}$$
(calculator answer)
$$= 1.88 \text{ moles O (correct answer)}$$

The ratio of 1.25 moles Fe/1.88 moles O must be changed to small whole numbers. This is done by dividing each by the smallest value:

$$\frac{1.25 \text{ moles Fe}}{1.25} = 1.00 \text{ moles Fe}$$
(calculator and correct answer)

$$\frac{1.88 \text{ moles O}}{1.25} = 1.504 \text{ moles O (calculator answer)}$$
$$= 1.50 \text{ moles O (correct answer)}$$

If these were all whole numbers (or could be rounded off to all whole numbers without introducing an error of more than 1%), we would now know the subscripts in the formula.

The Mole Concept and Chemical Formulas 119

When a small decimal number is obtained, as is the case here -- 1.50, such decimals must be "cleared". This is accomplished by multiplying all of the mole values by a small whole number. For the decimal 0.5 multiplication is by 2, for 0.33 or 0.67 (thirds) multiplication is by 3 and for 0.25 or 0.75 (fourths) multiplication is by four. Thus, multiplying by 2 we get:

1.00 moles Fe x 2 = 2.00 moles Fe

1.50 moles O x 2 = 3.00 moles O

The empirical formula, therefore, is Fe_2O_3.

9. <u>Calculate a molecular formula for a substance from its</u> <u>empirical formula and formula weight information</u> (section 7.8).

In order to determine the molecular (or true) formula, it is necessary to have information about the formula weight of the compound, as well as the percentage composition.

Example: Calculate the empirical formula and the molecular formula of a compound which contains 8.56 g carbon and 1.44 g hydrogen, and has a molecular weight of 42.1 amu.

Solution:

We find the empirical formula, as in the previous problem:

8.56 g C x $\left(\dfrac{1 \text{ mole C}}{12.0 \text{ g C}}\right)$ = 0.71333333 moles C (calculator answer)
= 0.713 moles C (correct answer)

1.44 g H x $\left(\dfrac{1 \text{ mole H}}{1.01 \text{ g H}}\right)$ = 1.4257426 moles H (calculator answer)
= 1.43 moles H (correct answer)

moles C = 0.713/0.713 = 1.00 moles C (calculator and correct answer)

moles H = 1.43/0.713 = 2.0056101 moles H (calculator answer)
= 2.01 moles H (correct answer)

The empirical formula is CH_2 (rounding 2.01 off to 2

involves an error of less than 1%, and so it is justified).

The molecular formula is either the same as the empirical formula ot it is a multiple of it.

If the "molecular weight" of the empirical formula is divided into the given molecular weight, the whole number answer tells how many times heavier the formula unit really is.

The "molecular weight" (or formula weight) of CH_2 =

$$
\begin{array}{ll}
1\ C = & 12.0\ \text{amu} \\
\underline{2\ H =} & \underline{2.0\ \text{amu}} \\
& 14.0\ \text{amu}
\end{array}
$$

The true molecular weight was given as 42.1 amu.

$$
\frac{42.1\ \cancel{\text{amu}}}{14.0\ \cancel{\text{amu}}} = 3.0071429\ \text{(calculator answer)}
$$
$$
= 3.00\ \text{(correct answer)}
$$

Therefore, the molecular formula is three times heavier than the empirical formula:

$3 \times CH_2 = C_3H_6$ (C_3H_6 is the molecular formula)

PROBLEM SET

1. Calculate the formula weight of each of the following compounds, to as many decimal places as possible.

   a) SO_3
   b) $Ca(NO_3)_2$

2. Calculate the percentage composition, to 2 decimal places, of:

   a) SO_3
   b) $Ca(NO_3)_2$

3. Analysis of a compound shows it contains 22.15 g Cr and 10.22 g O. Calculate:

   a) the percentage composition of the compound

b) the empirical formula of the compound

4. Calculate the number of atoms in:

    a) 125.3 moles of silver
    b) 8.323 grams of chromium

5. Calculate the number of molecules in:

    a) 12.34 moles of $C_{12}H_{22}O_{11}$
    b) 10.57 grams of CCl_4

6. Calculate the mass, in grams, of:

    a) 2.54 moles of N atoms

    b) 2.54 moles of N_2 molecules

    c) 50.0 moles of HNO_3 molecules

7. Determine which of the following pairs contain equal numbers of atoms:

    a) 1.000 grams of Pt and 1.062 grams of Pb
    b) 1.000 grams of Na and 1.155 grams of K

8. Determine which of the following pairs contain equal numbers of molecules:

    a) 5.000 grams of C_2H_5OH and 4.000 grams of CH_2O
    b) 1.545 grams of PH_3 and 5.472 grams of NCl_3

9. Calculate the number of moles of O atoms in 125.8 moles of $Al_2(SO_4)_3$.

10. Calculate the number of O atoms in 12.58 grams of $Al_2(SO_4)_3$.

11. Calculate the mass, in grams, of exactly 3 million atoms of gold, Au.

12. Calculate the mass, in grams, of one molecule of NH_3.

13. A compound contains 87.4% N and 12.6% H. Calculate:

    a) the empirical formula of this compound
    b) the molecular formula of this compound, if its molecular weight is 32.0 amu.

MULTIPLE CHOICE EXERCISES

1. To find the formula weight of a substance you should:

   a) add up the atomic weights of all the atoms present
   b) add up the atomic numbers of all the atoms present
   c) determine the number of protons present and multiply this number by 12
   d) pick up a single molecule of the substance (being careful not to squeeze it too hard), place it on the bathroom scale and weigh it

2. The formula weight of CO_2 is:

   a) 23.0 amu
   b) 30.0 amu
   c) 44.0 amu
   d) 48.0 amu

3. The formula weight of $Ca_3(PO_4)_2$ is:

   a) 87.0 amu
   b) 154 amu
   c) 279 amu
   d) 310 amu

4. The percentage of hydrogen (by mass) in the compound NH_3:

   a) 10.3%
   b) 17.8%
   c) 25.0%
   d) 75.0%

5. As a counting unit, a mole is equal to:

   a) 300 dozen objects
   b) 6.02×10^{23} objects
   c) one billion objects
   d) a grundle of objects

6. The number of atoms in 1.5 moles of Cu is:

   a) 2.0 atom
   b) 3.0×10^{23} atoms
   c) 6.0×10^{23} atoms
   d) 9.0×10^{23} atoms

7. The atomic weights of He and Be are 4.0 and 9.0 amu respectively. Which of the following statements is true?

   a) a mole of helium weighs the same as a mole of beryllium
   b) a mole of helium is heavier than a mole of beryllium
   c) a mole of beryllium is heavier than a mole of helium
   d) a mole of beryllium contains more molecules than a mole of helium

8. The molar mass for CO is:

   a) 15.0 g
   b) 17.3 g
   c) 28.0 g
   d) 53.4 g

9. 34.0 grams of NH_3 is equal to how many moles of NH_3?

   a) 0.50 moles
   b) 1.0 moles
   c) 1.5 moles
   d) 2.0 moles

10. Which of the following weights of copper and gold would contain the same number of atoms?

    a) 10 grams of copper and 10 grams of gold
    b) 29 grams of copper and 79 grams of gold
    c) 63.5 grams of copper and 197.0 grams of gold
    d) 197.0 grams of copper and 63.5 grams of gold

11. One mole of N_2O_4 contains:

    a) two atoms N
    b) two molecules N_2O_4
    c) four moles O
    d) six moles N_2O_4

12. Which of the following contains the largest number of atoms?

    a) 2 moles of CO_2

    b) 3 moles of N_2

    c) 6 moles of He

    d) 2 moles of NH_3

13. Which of the following is the correct set-up for the problem "How many atoms are there in 28.76 grams of C?"

a) $28.76 \text{ g C} \times \dfrac{1 \text{ mole C}}{12.01 \text{ g C}} \times \dfrac{1 \text{ atom C}}{6.02 \times 10^{23} \text{ moles C}}$

b) $28.76 \text{ g C} \times \dfrac{12.01 \text{ g C}}{1 \text{ mole C}} \times \dfrac{6.02 \times 10^{23} \text{ atoms C}}{1 \text{ mole C}}$

c) $28.76 \text{ g C} \times \dfrac{1 \text{ mole C}}{12.01 \text{ g C}} \times \dfrac{6.02 \times 10^{23} \text{ atoms C}}{1 \text{ mole C}}$

d) $28.76 \text{ g C} \times \dfrac{12.01 \text{ g C}}{1 \text{ mole C}} \times \dfrac{1 \text{ mole C}}{6.02 \times 10^{23} \text{ atoms C}}$

14. The set-up for the problem "What is the weight in grams of a single atom of gold?" that follows is correct, except for the replacement of some numbers in the conversion factors by letters. Which of the statements concerning the letters is correct?

$$1 \text{ atom Au} \times \dfrac{1 \text{ mole Au}}{A \text{ atoms Au}} \times \dfrac{B \text{ grams Au}}{C \text{ moles Au}}$$

a) A = 197.0
b) $B = 6.02 \times 10^{23}$
c) $C = 6.02 \times 10^{23}$
d) B = 197.0

15. Which of the following is the correct set-up for the problem "How many oxygen atoms are there in 4.0 moles of N_2O_5?"

a) $4.0 \text{ moles } N_2O_5 \times \dfrac{5 \text{ moles O}}{1 \text{ mole } N_2O_5} \times \dfrac{6.02 \times 10^{23} \text{ atoms O}}{5 \text{ moles O}}$

b) $4.0 \text{ moles } N_2O_5 \times \dfrac{5 \text{ moles O}}{1 \text{ mole } N_2O_5} \times \dfrac{6.02 \times 10^{23} \text{ atoms O}}{1 \text{ mole O}}$

c) $4.0 \text{ moles } N_2O_5 \times \dfrac{1 \text{ mole O}}{1 \text{ mole } N_2O_5} \times \dfrac{6.02 \times 10^{23} \text{ atoms O}}{5 \text{ moles O}}$

d) $4.0 \text{ moles } N_2O_5 \times \dfrac{1 \text{ mole O}}{1 \text{ mole } N_2O_5} \times \dfrac{6.02 \times 10^{23} \text{ atoms O}}{1 \text{ mole O}}$

16. Which of the following is the empirical formula of $C_6H_4Cl_2$?

   a) C_3H_2Cl
   b) $CHCl$
   c) C_2H_2Cl
   d) C_2HCl

17. If the empirical formula of a compound is CH_2 and the formula weight is 84, what is the molecular formula?

   a) C_2H_4
   b) C_3H_6
   c) C_4H_8
   d) C_6H_{12}

18. What is the empirical formula of a compound that contains 9.90% C, 58.7% Cl and 31.4% F?

   a) CCl_6F_3
   b) $C_2Cl_2F_2$
   c) CCl_2F_2
   d) C_4Cl_5F

19. One atom of helium (He) has a mass of how many grams?

   a) 4.00 g
   b) 6.64 x 10^{-22} g
   c) 6.64 x 10^{-23} g
   d) 6.64 x 10^{-24} g

20. 1.50 moles of Ag_2SO_4 has a mass of how many grams?

   a) 216 g
   b) 312 g
   c) 468 g
   d) 503 g

ANSWERS TO PROBLEMS SET

1. a) One formula unit of SO_3 contains 1 atom of S and 3
       atoms of O.

       S: 1 x 32.06   amu = 32.06   amu
       O: 3 x 15.9994 amu = 47.9982 amu

                          = 80.0582 amu (calculator answer)
                          = 80.06 amu (correct answer)

    b) One formula unit of $Ca(NO_3)_2$ contains 1 atom of Ca, 2
       atoms of N, and 6 atoms of O.

       Ca: 1 x 40.08   amu = 40.08   amu
        N: 2 x 14.0067 amu = 28.0134 amu
        O: 6 x 15.9994 amu = 95.9964 amu

                          = 164.0898 amu
                            (calculator answer)
                          = 164.09 amu (correct answer)

2. a) %S: $\dfrac{32.06 \text{ amu}}{80.06 \text{ amu}}$ x 100 = 40.044966 %S
                              (calculator answer)
                          = 40.04 %S (correct answer)

       %O: $\dfrac{48.00 \text{ amu}}{80.06 \text{ amu}}$ x 100 = 59.955034 %O
                              (calculator answer)
                          = 59.96 %O (correct answer)

    b) %Ca: $\dfrac{40.08 \text{ amu}}{164.1 \text{ amu}}$ x 100 = 24.424132 %Ca
                              (calculator answer)
                          = 24.42 %Ca (correct answer)

       %N: $\dfrac{28.01 \text{ amu}}{164.1 \text{ amu}}$ x 100 = 17.06886 %N (calculator answer)

                          = 17.07 %N (correct answer)

       %O: $\dfrac{96.00 \text{ amu}}{164.1 \text{ amu}}$ x 100 = 58.500914 %O
                              (calculator answer)
                          = 58.50% O (correct answer)

3. a) 22.15 g Cr + 10.22 g O = 32.37 g of compound
                              (calculator and
                               correct answer)

       %Cr: $\dfrac{22.15 \text{ g}}{32.37 \text{ g}}$ x 100 = 68.427556 %Cr
                              (calculator answer)
                          = 68.43% Cr (correct answer)

%O: $\dfrac{10.22 \text{ g}}{32.37 \text{ g}}$ x 100 = 31.572444 %O (calculator answer)

$\qquad\qquad\qquad\qquad$ = 31.57% O (correct answer)

b) 22.15 g Cr x$\left(\dfrac{1 \text{ mole Cr}}{52.00 \text{ g Cr}}\right)$ = 0.42596154 moles Cr

$\qquad\qquad\qquad\qquad\qquad$ (calculator answer)

$\qquad\qquad\qquad\qquad\qquad$ = 0.4260 moles Cr

$\qquad\qquad\qquad\qquad\qquad$ (correct answer)

10.22 g O x$\left(\dfrac{1 \text{ mole O}}{16.00 \text{ g O}}\right)$ = 0.63875 mole O

$\qquad\qquad\qquad\qquad\qquad$ (calculator answer)

$\qquad\qquad\qquad\qquad$ = 0.6388 mole O (correct answer)

Moles Cr = $\dfrac{0.4260}{0.4260}$ = 1.000 moles Cr

$\qquad\qquad\qquad$ (calculator and correct answer)

Moles O = $\dfrac{0.6388}{0.4260}$ = 1.4995305 moles O

$\qquad\qquad\qquad$ (calculator answer)

$\qquad\qquad\qquad$ = 1.500 moles O (correct answer)

Since these are not both whole numbers, and since 0.5 = 1/2, multiplying both of these values by 2 gives a whole number ratio.

1.000 mole Cr x 2 = 2.000 moles Cr

1.500 moles O x 2 = 3.000 moles O

The empirical formula is Cr_2O_3

4. a) 125.3 moles Ag x$\left(\dfrac{6.022 \times 10^{23} \text{ Ag atoms}}{1 \text{ mole Ag}}\right)$

$\qquad\qquad$ = 7.545566 x 10^{25} Ag atoms (calculator answer)

$\qquad\qquad$ = 7.546 x 10^{25} Ag atoms (correct answer)

b) 8.323 g Cr x$\left(\dfrac{1 \text{ mole Cr atoms}}{52.00 \text{ g Cr}}\right)x\left(\dfrac{6.022 \times 10^{23} \text{ Cr atoms}}{1 \text{ mole Cr atoms}}\right)$

$\qquad\qquad$ = 9.6386742 x 10^{22} Cr atoms (calculator answer)

$\qquad\qquad$ = 9.639 x 10^{22} Cr atoms (correct answer)

5. a) 12.34 moles $C_{12}H_{22}O_{11}$ x$\left(\dfrac{6.022 \times 10^{23} \text{ molecules}}{1 \text{ mole } C_{12}H_{22}O_{11}}\right)$

$\qquad\qquad$ = 7.431148 x 10^{24} molecules (calculator answer)

$\qquad\qquad$ = 7.431 x 10^{24} molecules (correct answer)

b) Molecular weight of CCl_4 is:

C: 1 x 12.01 amu = 12.01 amu
Cl: 4 x 35.45 amu =141.80 amu

 153.81 amu (calculator and
 correct answer)

$$10.57 \text{ g } CCl_4 \times \left(\frac{1 \text{ mole } CCl_4}{153.8 \text{ g } CCl_4}\right) \times \left(\frac{6.022 \times 10^{23} \text{ molecules } CCl_4}{1 \text{ mole } CCl_4}\right)$$

$= 4.1386567 \times 10^{22}$ molecules CCl_4 (calculator answer)

$= 4.139 \times 10^{22}$ molecules CCl_4 (correct answer)

6. a) 2.54 moles of N atoms $\times \left(\dfrac{14.0 \text{ g N}}{1 \text{ mole N atoms}}\right)$

 = 35.56 g N (calculator answer)
 = 35.6 g N (correct answer)

b) 2.54 moles of N_2 molecules $\times \left(\dfrac{28.0 \text{ g } N_2}{1 \text{ mole } N_2 \text{ molecules}}\right)$

 = 71.12 g N_2 (calculator answer)
 = 71.1 g N_2 (correct answer)

c) The molecular weight of HNO_3 is 63.0 amu.

50.0 moles $HNO_3 \times \left(\dfrac{63.0 \text{ g } HNO_3}{1 \text{ mole } HNO_3}\right)$ = 3150 g HNO_3
 (calculator and
 correct answer)

7. a) $\dfrac{\text{atomic weight Pt}}{\text{atomic weight Pb}} = \dfrac{195.1 \text{ amu}}{207.2 \text{ amu}}$ = 0.94160232
 (calculator answer)
 = 0.9416 (correct answer)

$\dfrac{\text{mass Pt}}{\text{mass Pb}} = \dfrac{1.000 \text{ g Pt}}{1.062 \text{ g Pb}}$ = 0.94161959 (calculator answer)
 = 0.9416 (correct answer)

The atomic weight ratio and the mass ratio are the
same, therefore equal numbers of atoms are present.

b) $\dfrac{\text{atomic weight Na}}{\text{atomic weight K}} = \dfrac{22.99 \text{ amu}}{39.10 \text{ amu}}$ = 0.58797954
 (calculator answer)

 = 0.5880 (correct answer)

$$\frac{\text{mass Na}}{\text{mass K}} = \frac{1.000 \text{ g Na}}{1.155 \text{ g K}} = 0.86580087 \text{ (calculator answer)}$$
$$= 0.8658 \text{ (correct answer)}$$

The atomic weight ratio and the mass ratio are not the same, therefore equal numbers of atoms are not present.

8. a) The molecular weight of C_2H_5OH is 46.07 amu and the molecular weight of CH_2O is 30.03 amu.

$$\frac{\text{molecular weight of } C_2H_5OH}{\text{molecular weight of } CH_2O} = \frac{46.07 \text{ amu}}{30.03 \text{ amu}}$$
$$= 1.5341325 \text{ (calculator answer)}$$
$$= 1.534 \text{ (correct answer)}$$

$$\frac{\text{mass of } C_2H_5OH}{\text{mass of } CH_2O} = \frac{5.000 \text{ g}}{4.000 \text{ g}} = 1.25 \text{ (calculator answer)}$$
$$= 1.250 \text{ (correct answer)}$$

The molecular weight ratio and the mass ratio are not the same, therefore equal numbers of molecules are not present.

b) The molecular weight of PH_3 is 33.99 amu and that of NCl_3 is 120.4 amu.

$$\frac{\text{molecular weight } PH_3}{\text{molecular weight } NCl_3} = \frac{33.99 \text{ amu}}{120.4 \text{ amu}} = 0.28230897$$
$$\text{(calculator answer)}$$
$$= 0.2823$$
$$\text{(correct answer)}$$

$$\frac{\text{mass } PH_3}{\text{mass } NCl_3} = \frac{1.545 \text{ g}}{5.472 \text{ g}} = 0.28234649 \text{ (calculator answer)}$$
$$= 0.2823 \text{ (correct answer)}$$

The molecular weight ratio and the mass ratio are the same, therefore equal numbers of molecules are present.

9. 125.8 $\text{moles Al}_2(SO_4)_3$ x $\left(\dfrac{12 \text{ moles O atoms}}{1 \text{ mole Al}_2(SO_4)_3}\right)$

x $\left(\dfrac{6.022 \times 10^{23} \text{ O atoms}}{1 \text{ mole O atoms}}\right)$ = 9.0908112 x 10^{26} O atoms
(calculator answer)

= 9.091 x 10^{26} O atoms
(correct answer)

10. The formula weight of $Al_2(SO_4)_3$ is 342.1 amu.

$$12.58 \text{ g } Al_2(SO_4)_3 \times \left(\frac{1 \text{ mole } Al_2(SO_4)_3}{342.1 \text{ g } Al_2(SO_4)_3}\right) \times \left(\frac{12 \text{ moles O atoms}}{1 \text{ mole } Al_2(SO_4)_3}\right)$$

$$\times \left(\frac{6.022 \times 10^{23} \text{ O atoms}}{1 \text{ mole O atoms}}\right) = 2.6573549 \times 10^{23} \text{ O atoms}$$
(calculator answer)

$$= 2.657 \times 10^{23} \text{ O atoms}$$

11. $$3.000 \times 10^6 \text{ Au atoms} \times \left(\frac{1 \text{ mole Au atoms}}{6.022 \times 10^{23} \text{ Au atoms}}\right)$$

$$\times \left(\frac{197.0 \text{ g Au}}{1 \text{ mole Au atoms}}\right) = 9.8140153 \times 10^{-16} \text{ g Au}$$
(calculator answer)

$$= 9.814 \times 10^{-16} \text{ g Au}$$
(correct answer)

12. $$1 \text{ molecule } NH_3 \times \left(\frac{1 \text{ mole } NH_3}{6.022 \times 10^{23} \text{ molecules } NH_3}\right)$$

$$\times \left(\frac{17.03 \text{ g } NH_3}{1 \text{ mole } NH_3}\right) = 2.8279641 \times 10^{-23} \text{ g } NH_3$$
(calculator answer)

$$= 2.828 \times 10^{-23} \text{ g } NH_3$$
(correct answer)

13. a) In 100.0 grams of this compound, there are 87.4 g N and 12.6 g H.

$$87.4 \text{ g N} \times \left(\frac{1 \text{ mole N}}{14.0 \text{ g N}}\right) = 6.2428571 \text{ moles N}$$
(calculator answer

$$= 6.24 \text{ moles N (correct answer)}$$

$$12.6 \text{ g H} \times \left(\frac{1 \text{ mole H}}{1.01 \text{ g H}}\right) = 12.475248 \text{ moles}$$
(calculator answer)

$$= 12.5 \text{ moles H (correct answer)}$$

moles of N: $\dfrac{6.24}{6.24} = 1.00$ mole N
(calculator and correct answer)

moles of H: $\dfrac{12.5}{6.24} = 2.0032051$ moles H
(calculator answer)
$$= 2.00 \text{ moles H (correct answer)}$$

The empirical formula of this compound is NH_2.

b) The empirical formula, calculated in part a), is NH_2. The formula weight, of the empirical formula NH_2, is

$$1(14.0)amu + 2(1.01)amu = 16.02 \text{ amu (calculator answer)}$$
$$= 16.0 \text{ amu (correct answer)}$$

The formula weight of the molecular formula is given as 32.0 amu.

$$\frac{32.0 \text{ amu}}{16.0 \text{ amu}} = 2 \text{ (calculator answer)}$$
$$= 2.00 \text{ (correct answer)}$$

The molecular formula is therefore 2 times larger than the empirical formula:

$$2 \times NH_2 = N_2H_4$$

ANSWERS TO MULTIPLE CHOICE EXERCISES

1.	a	11.	c
2.	c	12.	d
3.	d	13.	c
4.	b	14.	d
5.	b	15.	b
6.	d	16.	a
7.	c	17.	d
8.	c	18.	c
9.	d	19.	d
10.	c	20.	c

CHAPTER 8
Chemical Calculations II: Calculations Involving Chemical Equations

REVIEW OF CHAPTER OBJECTIVES

1. State the law of conservation of mass and describe how atomic theory accounts for this law (section 8.1).

    According to atomic theory, an atom is the smallest unit of an element which takes part in a chemical reaction. Atoms are neither created nor destroyed in ordinary chemical reactions, they are merely rearranged. Therefore, the same number of the same kinds of atoms are present before and after a chemical reaction. This generalization is known as the Law of Conservation of Mass. An alternative statement of this law is that the mass of the products equals the mass of the reactants in a chemical reaction. In the reaction:

    $$Zn + S \longrightarrow ZnS$$

    it is experimentally determined that 65.38 grams of zinc react with 32.07 grams of sulfur to form 97.45 grams of zinc sulfide. Since 65.38 + 32.07 = 97.45, these numbers are consistent with the law of conservation of mass.

2. Understand the conventions used in writing chemical equations (section 8.2).

    Chemical equations use symbols and formulas (instead of words) to describe the changes that occur in a chemical reaction. Conventions in writing chemical equations include:

    a) Reactants, the substances that a chemical reaction starts with, are written on the left side of the equation.

$$CuO + H_2 \longrightarrow$$

b) Products, the substances that are produced as a result of the chemical reaction, are written on the right side of the equation.

$$CuO + H_2 \longrightarrow H_2O + Cu$$

c) An arrow is used to separate the reactants from the products.

$$CuO + H_2 \longrightarrow H_2O + Cu$$

$$\text{reactants} \qquad \text{products}$$

d) Plus signs separate individual reactants, or products, from each other. The plus sign on the left is read as "reacts with" (or plus). The arrow reads "produces" or "yields" or "forms". The plus sign on the right reads "and" (or plus).

Stated in words the chemical equation:

$$CuO + H_2 \longrightarrow H_2O + Cu$$

reads "copper(II) oxide reacts with hydrogen to form water and copper."

e) A chemical equation must agree with experimental facts. This means that all formulas for reactants and products must be correct. Elements that exist as diatomic molecules are shown in molecular form. You should learn that the following seven elements are written as diatomic molecules when they appear in a chemical reaction as free elements:

$$H_2, O_2, N_2, F_2, Cl_2, Br_2, I_2$$

Several free elements exist as polyatomic molecules. The two you are most likely to encounter are:

$$P_4 \text{ and } S_8$$

3. <u>Recognize whether or not an equation is balanced, and balance chemical equations once formulas of reactants and products are given</u> (section 8.3).

A <u>balanced chemical equation</u> has the same number of atoms of each element on both sides of the equation.

An unbalanced equation is changed to a balanced equation by placing coefficients to the left of the chemical formulas in the equation. The coefficients tell how many formula units are needed. The subscripts in a correctly written formula can never be changed in the balancing process.

Unbalanced equation -- all formulas are correct:

$$Al + Cl_2 \longrightarrow AlCl_3$$

Balanced equation -- no subscripts have been changed; coefficients have been added:

$$2Al + 3Cl_2 \longrightarrow 2AlCl_3$$

Here is another example.

Unbalanced equation:

$$CO + O_2 \longrightarrow CO_2$$

Balanced equation:

$$2CO + O_2 \longrightarrow 2CO_2$$

The coefficient of one is not written in a balanced equation. It is understood, in the same way that the formula CO_2 means that one molecule of CO_2 contains one atom of carbon and two atoms of oxygen.

In the balanced equation: $2Al + 3Cl_2 \longrightarrow 2AlCl_3$
there are 2 atoms of Al and 6 atoms of Cl, on both sides of the equation.

In the balanced equation: $2CO + O_2 \longrightarrow 2CO_2$
there are 2 atoms of carbon and 4 atoms of oxygen on both sides of the equation.

Example: Balance the equation:

$$Al(OH)_3 + H_2SO_4 \longrightarrow Al_2(SO_4)_3 + H_2O$$

Solution:

Step 1: Examine the equation, and pick one element to balance first.

Calculations Involving Chemical Equations 135

It is always better to balance the elements that appear in only one reactant and one product first, before trying to balance any element appearing in more than one product or more than one reactant. Therefore, in this equation, it is best to leave O for last, because O is found in both reactants and both products. The H is left for next to last, because it is found in 2 reactants.

The best element to start with is Al. The number of Al atoms is set at 2 on the right side of the equation in the formula:

$$Al_2(SO_4)_3 \; .$$

2 Al atoms are indicated on the left side of the equation by placing a coefficient of 2 in front of the formula $Al(OH)_3$:

$$2Al(OH)_3 + H_2SO_4 \longrightarrow Al_2(SO_4)_3 + H_2O$$

Step 2: Pick a second element to balance.

The element S is the only other element found in only one reactant and one product. It is helpful to consider polyatomic ions as single entities, when they maintain their identity in the chemical reaction. Therefore, in this reaction the sulfate ion (SO_4^{-2}) can be considered as a unit. There are 3 sulfate ions on the right side of the equation in $Al_2(SO_4)_3$, so a coefficient of 3 is needed before the formula H_2SO_4 on the left side of the equation to show 3 sulfate ions on the left side of the equation:

$$2Al(OH)_3 + 3H_2SO_4 \longrightarrow Al_2(SO_4)_3 + H_2O$$

Step 3: Pick a third element to balance.

The next element to balance would be hydrogen, which is found in 2 reactants and one product. On the left side of the equation there are 6 H atoms in the $2Al(OH)_3$, and 6 H atoms in the $3H_2SO_4$, for a total of 12 H atoms altogether on the left side of the equation.

To show 12 H atoms on the right side of the equation, the coefficient of 6 must be placed before the formula of water, H_2O:

$$2Al(OH)_3 + 3H_2SO_4 \longrightarrow Al_2(SO_4)_3 + 6H_2O$$

So far, we have checked that there are 2 Al, 3 S and 12 H atoms on both sides. If the equation is now "balanced", the number of O on both sides must also be the same, so we count up the number of O atoms on each side of the equation.

Left side: 2 Al(OH)$_3$ 2 x 3 =   6 atoms O

           3 H$_2$SO$_4$ 3 x 4 = $\underline{12 \text{ atoms O}}$

                                      18 atoms O

Right side: 1 Al$_2$(SO$_4$)$_3$ 3 x 4 = 12 atoms O

            6 H$_2$O   6 x 1 = $\underline{\;6 \text{ atoms O}}$

                                      18 atoms O

Since the number of Al (2), the number of S (3), the number of H (12) and the number of O (18) are the same on both sides of the equation, the equation is balanced.

4. <u>Understand the special symbols used in chemical equations</u> (section 8.4).

In addition to the essential symbols:

      $\longrightarrow$ "to produce"

            and

      +   "plus" or "reacts with"

chemical equations often show optional symbols to convey added information. These symbols include those that tell the state of the substance:

(s)   solid

(ℓ)   liquid

(g)   gas

(aq) aqueous solution (substance is dissolved in water)

As an example, when these optional symbols are included in the equation in the previous exercise, it would appear as:

   2Al(OH)$_{3(s)}$ + 3H$_2$SO$_{4(aq)}$ $\longrightarrow$ Al$_2$(SO$_4$)$_{3(s)}$ + 6H$_2$O$_{(\ell)}$

This balanced equation shows that the reactants are solid

Calculations Involving Chemical Equations   137

Al(OH)$_3$ and an aqueous solution of H$_2$SO$_4$, and the products are solid aluminum sulfate and liquid water. (Neither Al(OH)$_3$ nor Al$_2$(SO$_4$)$_3$ are soluble in water, and H$_2$O is a liquid at ordinary room temperature.)

5. Interpret coefficients in balanced equations in terms of moles (section 8.5).

The coefficients in a balanced equation can be interpreted on a microscopic level and on a macroscopic level. Interpreted on a microscopic level, the equation:

$$2SO_2 + O_2 \longrightarrow 2SO_3$$

reads: 2 molecules of SO$_2$ react with 1 molecule of O$_2$ to form 2 molecules of SO$_3$.

On a macroscopic level the coefficients in a balanced equation tell the molar ratios between reactants consumed and products produced. Interpreted on a macroscopic level, the equation:

$$2SO_2 + O_2 \longrightarrow 2SO_3$$

reads: 2 moles of SO$_2$ react with 1 mole of O$_2$ to produce 2 moles of SO$_3$.

The coefficients in an equation can be used to generate conversion factors for use in solving problems involving chemical equations. In the equation:

$$2SO_2 + O_2 \longrightarrow 2SO_3$$

three mole-to-mole relationships are obtainable.

1. 2 moles of SO$_2$ react with 1 mole of O$_2$
2. 2 moles of SO$_2$ produce 2 moles of SO$_3$
3. 1 mole of O$_2$ produces 2 moles of SO$_3$

From these mole-to-mole ratios, six conversion factors can be written:

from 1. $\dfrac{2 \text{ moles } SO_2}{1 \text{ mole } O_2}$ and $\dfrac{1 \text{ mole } O_2}{2 \text{ moles } SO_2}$

from 2. $\dfrac{2 \text{ moles } SO_2}{2 \text{ moles } SO_3}$ and $\dfrac{2 \text{ moles } SO_3}{2 \text{ moles } SO_2}$

from 3. $\dfrac{1 \text{ mole } O_2}{2 \text{ moles } SO_3}$ and $\dfrac{2 \text{ moles } SO_3}{1 \text{ mole } O_2}$

Example: Given the equation:

$$2C_6H_{14} + 19 \ O_2 \longrightarrow 12CO_2 + 14H_2O$$

find the following:

a) the moles of O_2 needed to react with 15.5 moles of C_6H_{14}.

b) the moles of H_2O produced from 2.50 moles of O_2.

Solution:

Each question involves a one-step mole-to-mole calculation.

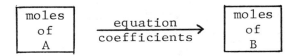

a) 15.5 moles C_6H_{14} = ? moles O_2

Since the balanced equation states that 2 moles of C_6H_{14} react with 19 moles of O_2, the 2 conversion factors are:

$\dfrac{2 \text{ moles } C_6H_{14}}{19 \text{ moles } O_2}$ and $\dfrac{19 \text{ moles } O_2}{2 \text{ moles } C_6H_{14}}$

The second factor is used so that the unit "moles of C_6H_{14}" will cancel (in numerator and denominator).

$$15.5 \ \cancel{\text{moles } C_6H_{14}} \ \times \left(\frac{19 \text{ moles } O_2}{2 \ \cancel{\text{moles } C_6H_{14}}} \right)$$

$$= 147.25 \text{ moles } O_2 \text{ (calculator answer)}$$

$$= 147 \text{ moles } O_2 \text{ (correct answer)}$$

b) 2.50 moles O_2 = ? moles H_2O

Since the balanced equation states that 19 moles of O_2 produce 14 moles of H_2O, the conversion factors are:

$$\frac{19 \text{ moles } O_2}{14 \text{ moles } H_2O} \quad \text{and} \quad \frac{14 \text{ moles } H_2O}{19 \text{ moles } O_2}$$

Using the second of the two conversion factors we get:

$$2.50 \text{ moles } O_2 \times \left(\frac{14 \text{ moles } H_2O}{19 \text{ moles } O_2} \right)$$

$$= 1.8421053 \text{ moles } H_2O \text{ (calculator answer)}$$

$$= 1.84 \text{ moles } H_2O \text{ (correct answer)}$$

In all problems relating molar quantities of reactants to products, or vice versa, the conversion factor needed is determined by the mole-to-mole ratio shown in the balanced equation. For this reason, a correctly balanced equation is a necessity.

6. <u>Use a balanced chemical equation and other appropriate information to calculate the quantities of reactants consumed and/or products produced in a chemical reaction</u> (section 8.6).

In a typical chemical equation-based calculation, information is given about one reactant or product of the reaction (which we can call "A"), and information needs to be calculated about another reactant or product (which we can call "B"). The quantity of either one may be in moles or grams or particles. The following diagram, which is also given on page 185 of the textbook, shows how the needed quantities are related.

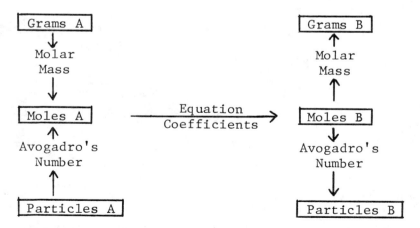

Two examples of the use of this diagram are as follows:

Example: Given the equation: $2Na + S \longrightarrow Na_2S$, find how many atoms of Na will react with 2.50 grams of S?

Solution: This is a "grams A-to-particles B" problem.

2.50 g S = ? atoms Na

The pathway obtained from the diagram is:

g S $\longrightarrow$ moles S $\longrightarrow$ moles Na $\longrightarrow$ atoms Na

The set-up, using dimensional analysis, is:

$$2.50 \; \cancel{g \; S} \times \left(\frac{1 \; \cancel{mole \; S}}{32.1 \; \cancel{g \; S}}\right) \times \left(\frac{2 \; \cancel{moles \; Na}}{1 \; \cancel{mole \; S}}\right) \times \left(\frac{6.02 \times 10^{23} \; atoms \; Na}{1 \; \cancel{mole \; Na}}\right)$$

$$= 9.376947 \times 10^{22} \; atoms \; Na \; (calculator \; answer)$$

$$= 9.38 \times 10^{22} \; atoms \; Na \; (correct \; answer)$$

Note that the solution for any calculation involving a chemical equation:

1) starts with a balanced equation

2) has a conversion factor somewhere in the solution that states the mole-to-mole ratio of the two substances being compared. This mole-to-mole ratio corresponds to the coefficients in the balanced equation which are in front of the substances being compared (and, when there is no coefficient shown, the number one is implied).

3) the measured quantities (such as grams of product or reactant) determines the number of significant figures in the answer. The mole-to-mole ratio is exact, and does not limit the number of significant figures in the answer.

Example: Given the equation:

$$2Al(OH)_3 + 3H_2SO_4 \longrightarrow Al_2(SO_4)_3 + 6H_2O$$

find how many grams of $Al_2(SO_4)_3$ are produced from 2.93 grams of $Al(OH)_3$.

Solution:

This is a "grams A-to-grams B" problem.   The pathway, obtained from the diagram, is:

g Al(OH)$_3$ ⟶ moles Al(OH)$_3$ ⟶ moles Al$_2$(SO$_4$)$_3$ ⟶ g Al$_2$(SO$_4$)$_3$

The set-up, using dimensional analysis, involves the following conversion factors:

$$2.93 \text{ g Al(OH)}_3 \times \frac{1 \text{ mole Al(OH)}_3}{78.0 \text{ g Al(OH)}_3} \times \frac{1 \text{ mole Al}_2(\text{SO}_4)_3}{2 \text{ moles Al(OH)}_3}$$

$$\times \frac{342 \text{ g Al}_2(\text{SO}_4)_3}{1 \text{ mole Al}_2(\text{SO}_4)_3} = 6.4234615 \text{ g Al}_2\text{SO}_4)_3$$

(calculator answer)

$$= 6.42 \text{ g Al}_2(\text{SO}_4)_3$$

(correct answer)

7. <u>Determine the limiting reactant in a chemical reaction given the mass of two or more reactants and the balanced chemical equation for the reaction</u> (section 8.7).

So far we have worked with situations in which reactants were present in the exact molar ratios shown in the balanced chemical equation. However, often one reactant is present in excess, perhaps to make a reaction go faster or to assure that all of an expensive reactant is completely used. When a reactant is present in excess, the excess will not react after all of the other reactant is used up. Therefore, the reactant which is <u>NOT</u> in excess limits the amount of product formed. The <u>limiting reactant</u> is the reactant which is consumed completely, while an excess of the other reactant(s) remain unused.

<u>Whenever you are given the quantities of two or more reactants in a chemical reaction it is necessary to determine which of the given quantities is the limiting reactant.</u> The amount of product formed must be calculated from the quantity of the limiting reactant.

Example: Under appropriate conditions sodium (Na) and sulfur (S) will react as shown by the equation:

2Na + S ⟶ Na$_2$S

If 5.00 grams of Na and 3.00 grams of S are mixed, which is the limiting reactant?

Solution:

We first determine how many moles of each reactant are actually present. This will involve simple "gram-to-mole" calculations.

142   Calculations Involving Chemical Equations

$$5.00 \text{ g Na} \times \left(\frac{1 \text{ mole Na}}{23.0 \text{ g Na}}\right) = 0.2173913 \text{ mole Na}$$
$$\text{(calculator answer)}$$
$$= 0.217 \text{ mole Na (correct answer)}$$

$$3.00 \text{ g S} \times \left(\frac{1 \text{ mole S}}{32.1 \text{ g S}}\right) = 0.09345794 \text{ mole S}$$
$$\text{(calculator answer)}$$
$$= 0.0935 \text{ mole S (correct answer)}$$

Then we determine how many moles of product Na_2S we can get from each of the previously determined number of moles of reactant. These calculations will be "mole-to-mole" calculations. The limiting reactant is the reactant that will produce the least number of moles of Na_2S.

$$0.217 \text{ moles Na} \times \left(\frac{1 \text{ mole Na}_2\text{S}}{2 \text{ moles Na}}\right) = 0.1085 \text{ moles Na}_2\text{S}$$
$$\text{(calculator answer)}$$
$$= 0.108 \text{ moles Na}_2\text{S}$$
$$\text{(correct answer)}$$

$$0.0935 \text{ moles S} \times \left(\frac{1 \text{ mole Na}_2\text{S}}{1 \text{ mole S}}\right) = 0.0935 \text{ moles Na}_2\text{S}$$
$$\text{(calculator and}$$
$$\text{correct answer)}$$

Thus, S is the limiting reactant.

Once the limiting reactant (which is sometimes called the limiting reagent) is determined, the amount of any product obtainable from the reactants can be calculated using the quantity of the limiting reactant present. The following exercise illustrates this:

Example: For the reaction:

$$2AgNO_3 + CaCl_2 \longrightarrow Ca(NO_3)_2 + 2AgCl$$

how many grams of AgCl are produced when 2.55 grams of $AgNO_3$ and 0.958 grams of $CaCl_2$ are mixed?

Solution:

First, the limiting reactant must be determined.

$$2.55 \text{ g AgNO}_3 \times \left(\frac{1 \text{ mole AgNO}_3}{170 \text{ g AgNO}_3}\right) = 0.015 \text{ mole AgNO}_3$$
$$\text{(calculator answer)}$$
$$= 0.0150 \text{ mole AgNO}_3$$
$$\text{(correct answer)}$$

Calculations Involving Chemical Equations 143

$$0.958 \text{ g } \cancel{CaCl_2} \times \left(\frac{1 \text{ mole } CaCl_2}{111 \text{ g } \cancel{CaCl_2}} \right) = 0.00863063 \text{ mole } CaCl_2$$

(calculator answer)

$$= 0.00863 \text{ mole } CaCl_2$$

(correct answer)

Next we determine how many moles of one of the products (we will arbitrarily select AgCl) we can produce from each of the reactants. The calculations will be mole-to-mole calculations that involve the use of the coefficients in the balanced equation in the conversion factor.

$$0.0150 \cancel{\text{mole } AgNO_3} \times \left(\frac{2 \text{ moles } AgCl}{2 \cancel{\text{moles } AgNO_3}} \right) = 0.0150 \text{ mole } AgCl$$

(calculator and correct answer)

$$0.00863 \cancel{\text{mole } CaCl_2} \times \left(\frac{2 \text{ moles } AgCl}{1 \cancel{\text{mole } CaCl_2}} \right) = 0.01726 \text{ moles } AgCl$$

(calculator answer)

$$= 0.0173 \text{ moles } AgCl$$

(correct answer)

$AgNO_3$, since it produces the least number of moles of AgCl, will be the limiting reactant.

Now we use the moles of AgCl produced by the limiting reactant to calculate the grams of AgCl produced.

$$0.0150 \cancel{\text{moles } AgCl} \times \left(\frac{143 \text{ g } AgCl}{1 \cancel{\text{mole } AgCl}} \right) = 2.145 \text{ g } AgCl$$

(calculator answer)

$$= 2.14 \text{ g } AgCl$$

(correct answer)

8. <u>Calculate the percent yield in a chemical reaction from appropriate data</u> (section 8.8).

The actual amount of product obtained in a chemical reaction is often far less than the calculated theoretical value. The reasons include loss of some of the product during isolation, or side reactions not shown in the equation.

The amount of product lost is indicated in terms of percent yield, a measure of the ratio of the actual yield and the theoretical yield.

Percent Yield = $\dfrac{\text{actual yield}}{\text{theoretical yield}} \times 100$

The <u>actual yield</u> is the amount of product actually obtained at the end of the experiment. It cannot be calculated. It must be experimentally determined.

The <u>theoretical yield</u> is the maximum amount of product that can be formed from the reactants if no losses of any kind occur. It is always a calculated number.

Example: If, for the reaction:

$$3AgNO_3 + Na_3PO_4 \longrightarrow Ag_3PO_4 + 3NaNO_3$$

one starts with 2.55 g $AgNO_3$ and excess Na_3PO_4, and recovers 1.92 g Ag_3PO_4, what is the percent yield of Ag_3PO_4?

Solution:

First we calculate the theoretical yield. This will involve a "grams A-to-grams B" calculation.

$$\text{g } AgNO_3 \longrightarrow \text{mole } AgNO_3 \longrightarrow \text{mole } Ag_3PO_4 \longrightarrow \text{g } Ag_3PO_4$$

$$2.55 \text{ g } AgNO_3 \times \left(\frac{1 \text{ mole } AgNO_3}{170 \text{ g } AgNO_3}\right) \times \left(\frac{1 \text{ mole } Ag_3PO_4}{3 \text{ mole } AgNO_3}\right)$$

$$\times \left(\frac{419 \text{ g } Ag_3PO_4}{1 \text{ mole } Ag_3PO_4}\right) = 2.095 \text{ g } Ag_3PO_4 \text{ (calculator answer)}$$
$$= 2.10 \text{ g } Ag_3PO_4 \text{ (correct answer)}$$

The actual yield is given as 1.92 g Ag_3PO_4. Thus,

Percent Yield = $\dfrac{\text{actual yield}}{\text{theoretical yield}} \times 100$

$$= \frac{1.92 \text{ g}}{2.10 \text{ g}} \times 100 = 91.428571 \text{ \%}$$
$$\text{(calculator answer)}$$
$$= 91.4\% \text{ (correct answer)}$$

9. OPTIONAL. <u>Calculate from appropriate data quantities of reactants consumed or products produced in chemical systems involving two or more simultaneous reactions or a series of consecutive reactions</u> (section 8.9).

In solving problems involving simultaneous reactions the usual procedure is to carry out parallel calculations on each of the simultaneous reactions. The individual answers obtained are then combined to give the "total"

Calculations Involving Chemical Equations   **145**

answer.

Example: An aluminum-zinc alloy reacts with hydrochloric acid, forming hydrogen gas according to the following equations:

$$2Al + 6HCl \longrightarrow 2AlCl_3 + 3H_2$$

$$Zn + 2HCl \longrightarrow ZnCl_2 + H_2$$

If 5.00 grams of the alloy, which is 25.5% Al and 74.5% Zn, reacts with hydrochloric acid, what weight of hydrogen gas forms?

Solution:

We first calculate the mass of each component of the alloy:

$$5.00 \text{ g alloy} \times \left(\frac{25.5 \text{ g Al}}{100 \text{ g alloy}}\right) = 1.275 \text{ g Al} \quad \text{(calculator answer)}$$
$$= 1.28 \text{ g Al} \quad \text{(correct answer)}$$

$$5.00 \text{ g alloy} \times \frac{74.5 \text{ g Zn}}{100 \text{ g alloy}} = 3.725 \text{ g Zn} \quad \text{(calculator answer)}$$
$$= 3.72 \text{ g Zn} \quad \text{(correct answer)}$$

Then we calculate the mass of hydrogen produced from each of the components of the alloy. This will involve two "grams A-to-grams B" type calculations.

Production of H_2 from Al is governed by the equation:

$$2Al + 6HCl \longrightarrow 2AlCl_3 + 3H_2$$

The dimensional analysis set-up is:

$$1.28 \text{ g Al} \times \left(\frac{1 \text{ mole Al}}{27.0 \text{ g Al}}\right) \times \left(\frac{3 \text{ moles } H_2}{2 \text{ moles Al}}\right) \times \left(\frac{2.02 \text{ g } H_2}{1 \text{ mole } H_2}\right)$$

$$= 0.14364444 \text{ g } H_2 \text{ (calculator answer)}$$
$$= 0.144 \text{ g } H_2 \text{ (correct answer)}$$

Production of H_2 from Zn is governed by the equation:

$$Zn + 2HCl \longrightarrow ZnCl_2 + H_2$$

The dimensional analysis set-up is:

$$3.72 \text{ g Zn } \times \left(\frac{1 \text{ mole Zn}}{65.4 \text{ g Zn}}\right) \times \left(\frac{1 \text{ mole H}_2}{1 \text{ mole Zn}}\right) \times \left(\frac{2.02 \text{ g H}_2}{1 \text{ mole H}_2}\right)$$

$$= 0.11482183 \text{ g H}_2 \text{ (calculator answer)}$$

$$= 0.115 \text{ g H}_2 \text{ (correct answer)}$$

Finally we add the masses of hydrogen obtained from the separate reactions to obtain our final answer.

0.144 g H_2 (from Al)

+ 0.115 g H_2 (from Zn)
_____

0.259 g H_2 (calculator and correct answer)

In a set of consecutive reactions one of the products from the first reaction is a reactant for the second reaction, one of the products from the second reaction is reactant for the third reaction and so on. These substances which "tie" the equations together are the "key substances" from a calculational point of view.

Example: An early industrial method of preparation of sulfuric acid (H_2SO_4) involved the following sequence of reactions:

$$S + O_2 \longrightarrow SO_2$$
$$2SO_2 + O_2 \longrightarrow 2SO_3$$
$$SO_3 + H_2O \longrightarrow H_2SO_4$$

If an excess of O_2 and H_2O are used, how many grams of H_2SO_4 are produced by 185 grams of S?

Solution:

The key substances in the reactions are the sulfur-containing species: S, SO_2, SO_3 and H_2SO_4.

$$S \xrightarrow{\text{reaction \#1}} SO_2 \xrightarrow{\text{reaction \#2}} SO_3 \xrightarrow{\text{reaction \#3}} H_2SO_4$$

The problem is a grams-to-grams problem with extra mole-mole intermediate steps.

185 g S = ? g H_2SO_4

The pathway is:

$$g\ S \longrightarrow moles\ S \longrightarrow moles\ SO_2 \longrightarrow moles\ SO_3$$

$$moles\ H_2SO_4 \longrightarrow g\ H_2SO_4$$

The equation coefficients in the first equation are the basis for the conversion factor effecting the moles S to moles SO_2 change. Similarly, the equation coefficients in the second and third reaction are used, respectively, in going from moles SO_2 to moles SO_3 and from moles SO_3 to moles H_2SO_4.

The set-up, using dimensional analysis, is:

$$185\ \cancel{g\ S} \times \left(\frac{1\ \cancel{mole\ S}}{32.1\ \cancel{g\ S}}\right) \times \left(\frac{1\ \cancel{mole\ SO_2}}{1\ \cancel{mole\ S}}\right) \times \left(\frac{2\ \cancel{moles\ SO_3}}{2\ \cancel{moles\ SO_2}}\right)$$

$$\times \left(\frac{1\ \cancel{mole\ H_2SO_4}}{1\ \cancel{mole\ SO_3}}\right) \times \left(\frac{98.1\ g\ H_2SO_4}{1\ \cancel{mole\ H_2SO_4}}\right)$$

$$= 565.37383\ g\ H_2SO_4\ \text{(calculator answer)}$$

$$= 565\ g\ H_2SO_4\ \text{(correct answer)}$$

This problem can also be solved in three separate steps:

Step 1: How many grams of SO_2 are produced from 185 g S?

Step 2: How many grams of SO_3 are produced from the SO_2 (in step 1)?

Step 3: How many grams of H_2SO_4 are produced from the SO_3 (in step 2)?

The answer, determined one step at a time, would be the same as the one calculated in the one-step method shown above.

PROBLEM SET

1.   Aluminum and fluorine react according to the equation:

$$2Al + 3F_2 \longrightarrow 2AlF_3$$

If 18.0 grams of Al react completely with 38.0 grams of F_2 how many grams of AlF_3 form?

2. List the seven elements which are written as diatomic molecules when they appear as free elements in equations.

3. Balance the following equations:

a) $N_2 + H_2 \longrightarrow NH_3$

b) $C + O_2 \longrightarrow CO$

c) $CO + O_2 \longrightarrow CO_2$

d) $C_2H_4 + O_2 \longrightarrow CO_2 + H_2O$

e) $C_2H_6 + O_2 \longrightarrow CO_2 + H_2O$

f) $H_2O_2 \longrightarrow H_2O + O_2$

g) $Na_2CO_3 + HCl \longrightarrow NaCl + H_2O + CO_2$

h) $Mg(NO_3)_2 + H_3PO_4 \longrightarrow Mg_3(PO_4)_2 + HNO_3$

i) $FeCl_3 + KOH \longrightarrow Fe(OH)_3 + KCl$

j) $Al(ClO_3)_3 \longrightarrow AlCl_3 + O_2$

4. Write the twelve mole-to-mole conversion factors that can be derived from the balanced equation:

$$2C_6H_{14} + 19\,O_2 \longrightarrow 12CO_2 + 14H_2O$$

5. How many moles of CO_2 are produced when 1.25 moles of the first reactant reacts with an excess of the second reactant listed?

a) $2CO + O_2 \longrightarrow 2CO_2$

b) $Ca(HCO_3)_2 + H_2SO_4 \longrightarrow CaSO_4 + 2H_2O + 2CO_2$

c) $2C_6H_{14} + 19\,O_2 \longrightarrow 12CO_2 + 14H_2O$

6. Given the equation: $C_{25}H_{52} + 38\,O_2 \longrightarrow 25\,CO_2 + 26H_2O$ answer the following:

a) 5.48 g $C_{25}H_{52}$ produces how many g CO_2?

b) 15.7 g O_2 reacts with how many g $C_{25}H_{52}$?

c) 5.48 moles O_2 produce how many g CO_2?

d) 1.57 g $C_{25}H_{52}$ produce how many molecules of H_2O?

e) How many g $C_{25}H_{52}$ are needed to produce 6.02×10^{20} molecules of CO_2?

7. Copper (Cu) and nitric acid (HNO_3) react as shown by the equation

$$Cu + 4HNO_3 \longrightarrow Cu(NO_3)_2 + 2H_2O + 2NO_2$$

If 5.86 g Cu and 23.1 g HNO_3 are mixed,
a) which reactant is the limiting reactant?
b) what mass of H_2O is formed?
c) how many molecules of NO_2 are formed?
d) how many grams of excess reactant is left over after the reaction is complete?

8. Under appropriate conditions $Fe(NO_3)_3$ and NH_4OH react to produce $Fe(OH)_3$ as shown by the following equation:

$$Fe(NO_3)_3 + 3NH_4OH \longrightarrow Fe(OH)_3 + 3NH_4NO_3$$

In a certain experiment 5.26 g of $Fe(OH)_3$ was produced from 12.5 g of $Fe(NO_3)_3$.

a) what is the theoretical yield of $Fe(OH)_3$?
b) what is the percent yield of $Fe(OH)_3$?

Optional

9. A mixture is 15.9% Ag_2O and 84.1% HgO. When this mixture is heated the oxides decompose completely to produce free elements, as shown by the equations:

$$2Ag_2O \longrightarrow 4Ag + O_2$$

$$2HgO \longrightarrow 2Hg + O_2$$

How many grams of O_2 are produced when 86.3 grams of the mixture decomposes?

10. How many grams of CO_2 can be obtained from 17.0 grams of $NaNO_3$ using the following two-step process?

$$2NaNO_3 \longrightarrow 2NaNO_2 + O_2$$

$$2CO + O_2 \longrightarrow 2CO_2$$

MULTIPLE CHOICE EXERCISES

1. Consider a reaction in which A and B are reactants and C and D are products. If 8 grams of A completely reacts with 11 grams of B to produce 5 grams of C, how many grams of D will be produced?

a) 5 g   c) 14 g
b) 9 g   d) 19 g

2. In a balanced chemical equation:

   a) the number of products must equal the number of reactants
   b) the sum of the coefficients on each side must be equal
   c) the sum of the subscripts on each side must be equal
   d) the total number of atoms on each side must be equal

3. Which of the following gases is written as a diatomic molecule in a chemical equation:

   a) helium
   b) nitrogen
   c) neon
   d) argon

4. Which of the following equations is balanced:

   a) $2Al + 3HCl \longrightarrow 2AlCl_3 + 3H_2$
   b) $2Na + H_2O \longrightarrow 2NaOH + H_2$
   c) $2KClO_3 \longrightarrow 2KCl + 3 O_2$
   d) $3SO_2 + 2 O_2 \longrightarrow 3SO_3$

5. The equation $Al + MnO_2 \longrightarrow Mn + Al_2O_3$, when correctly balanced, appears as:

   a) $2Al + 2MnO_2 \longrightarrow 2Mn + Al_2O_3$
   b) $4Al + 3MnO_2 \longrightarrow 3Mn + 2Al_2O_3$
   c) $2Al + MnO_2 \longrightarrow Mn + Al_2O_3$
   d) $6Al + 4MnO_2 \longrightarrow 4Mn + 3Al_2O_3$

6. When the equation $N_2O \longrightarrow N_2 + O_2$ is correctly balanced, what coefficient is in front of N_2O?

   a) 1
   b) 2
   c) 3
   d) 4

7. When the equation $S_8 + O_2 + H_2O \longrightarrow H_2SO_4$ is correctly balanced, what coefficient is in front of O_2?

   a) 4
   b) 6
   c) 8
   d) 12

8. In the reaction $Zn_{(s)} + 2HCl_{(aq)} \longrightarrow H_{2(g)} + ZnCl_{2(aq)}$, the $ZnCl_2$ is:

a) a reactant
b) in solution
c) a gas
d) a solid

9. Which one of the following conversion factors is not consistent with the equation:

$$2Cu(NO_3)_2 \longrightarrow 4NO_2 + O_2 + 2CuO$$

a) $\dfrac{2 \text{ moles } CuO}{4 \text{ moles } NO_2}$

b) $\dfrac{1 \text{ mole } O_2}{2 \text{ moles } Cu(NO_3)_2}$

c) $\dfrac{2 \text{ moles } Cu(NO_3)_2}{2 \text{ moles } CuO}$

d) $\dfrac{4 \text{ moles } NO_2}{2 \text{ moles } O_2}$

10. Which of the following statements is not consistent with the equation:

$$2CO + O_2 \longrightarrow 2CO_2$$

a) 2 moles CO plus 1 mole of O_2 react to produce 2 moles CO_2

b) 1/2 mole CO plus 1/4 mole of O_2 react to produce 1/2 mole CO_2

c) 6 moles CO plus 2 moles of O_2 react to produce 4 moles CO_2

d) 1 mole CO plus 1/2 mole of O_2 react to produce 1 mole CO_2

11. How many moles of NH_3 are needed to produce 6 moles of N_2 according to the following reaction:

$$4NH_3 + 3 O_2 \longrightarrow 2N_2 + 6H_2O$$

a) 2 moles
b) 4 moles
c) 8 moles
d) 12 moles

12. The correct set-up for the problem "How many grams of Fe_2O_3 (MW = 159.6 amu) are needed to completely react with 14.0 g of CO according to the following reaction?" is:

$$3Fe_2O_3 + CO \longrightarrow 2Fe_3O_4 + CO_2$$

a) 14.0 g CO x $\left(\dfrac{1 \text{ mole CO}}{28.0 \text{ g CO}}\right)$ x $\left(\dfrac{3 \text{ moles } Fe_2O_3}{1 \text{ mole CO}}\right)$

x $\left(\dfrac{159.6 \text{ g } Fe_2O_3}{1 \text{ mole } Fe_2O_3}\right)$

b) 14.0 g CO x $\left(\dfrac{1 \text{ mole CO}}{28.0 \text{ g CO}}\right)$ x $\left(\dfrac{1 \text{ mole } Fe_2O_3}{1 \text{ mole CO}}\right)$

x $\left(\dfrac{159.6 \text{ g } Fe_2O_3}{3 \text{ moles } Fe_2O_3}\right)$

c) 14.0 g CO x $\left(\dfrac{3 \text{ moles } Fe_2O_3}{1 \text{ mole CO}}\right)$ x $\left(\dfrac{159.6 \text{ g } Fe_2O_3}{1 \text{ mole } Fe_2O_3}\right)$

d) 14.0 g CO x $\left(\dfrac{1 \text{ mole } Fe_2O_3}{28 \text{ moles CO}}\right)$ x $\left(\dfrac{159.6 \text{ g } Fe_2O_3}{3 \text{ moles } Fe_2O_3}\right)$

13. The "set-up" for the problem "How many grams of CO_2 will be produced in the reaction:

$$CO + NO_2 \longrightarrow CO_2 + NO$$

if 38.0 g of NO are produced?" which follows is correct except some parts of the conversion factors have been replaced by letters. Which of the statements concerning the letters is correct?

38.0 g NO x $\left(\dfrac{A}{30.0 \text{ g NO}}\right)$ x $\left(\dfrac{B}{C}\right)$ x $\left(\dfrac{44.0 \text{ g } CO_2}{D}\right)$

a) A = 2 moles CO_2
b) B = 1 mole CO_2
c) C = 1 mole CO_2
d) D = 2 moles CO_2

Calculations Involving Chemical Equations   153

14. How many moles of oxygen (O_2) are produced by the decomposition of 12.26 g of $KClO_3$ (MW = 122.6 amu) according to the equation:

$$2KClO_3 \longrightarrow 2KCl + 3\ O_2 \ ?$$

a) 0.075 moles
b) 0.15 moles
c) 4.8 moles
d) 9.6 moles

Questions 15 through 17 pertain to the following chemical reaction:

$$3A + 2B + 6C \longrightarrow 4D + E$$

15. If a mixture involving 3 moles of A, 5 moles of B and an unlimited amount of C is allowed to react, then

a) A will be the limiting reactant
b) B will be the limiting reactant
c) C will be the limiting reactant
d) 2 of the 3 reactants may be considered limiting

16. If a mixture involving 3.0 moles of A, an unlimited amount of B and 6.0 moles of C is allowed to react, then

a) A will be the limiting reactant
b) B will be the limiting reactant
c) C will be the limiting reactant
d) 2 of the 3 reactants may be considered limiting

17. If a mixture involving 3 moles of A, 4 moles of B and 8 moles of C is allowed to react, then

a) A will be the limiting reactant
b) B will be the limiting reactant
c) C will be the limiting reactant
d) 2 of the 3 reactants will be limiting

18. Ammonia (NH_3) can be produced from the reaction

$$N_2 + 3H_2 \longrightarrow 2NH_3$$

How many moles of NH_3 can be produced from 14.0 g N_2 and 3.0 g H_2?

a) 0.67 moles
b) 3.0 moles
c) 14.0 moles
d) 17.0 moles

154 Calculations Involving Chemical Equations

19. Which of the following types of yields must always be determined experimentally?

    a) percentage yield
    b) actual yield
    c) theoretical yield
    d) interest yield

20. In an experiment 1 mole of CO_2 was produced from 4 moles of HCl and an excess of Na_2CO_3 according to the reaction

$$2HCl + Na_2CO_3 \longrightarrow 2NaCl + CO_2 + H_2O$$

The percentage yield of the reaction would be:

    a) 25%
    b) 50%
    c) 75%
    d) 100%

Optional

21. A mixture contains 1 mole of Ag_2O and 3 moles of HgO. When this mixture is heated both oxides decompose completely to produce free elements as shown by the equations:

$$2Ag_2O \longrightarrow 4Ag + O_2$$

$$2HgO \longrightarrow 2Hg + O_2$$

How many moles of O_2 are produced?

    a) 0.50 mole
    b) 1.5 moles
    c) 2.0 moles
    d) 4.0 moles

22. Given the set of consecutive equations:

$$2A + 3B \longrightarrow C + 2D$$

$$4D + E \longrightarrow 5F + 2G$$

How many moles of B are needed to produce 8.00 moles of G?

    a) 2.67
    b) 10.7
    c) 12.0
    d) 24.0

Calculations Involving Chemical Equations 155

SOLUTIONS TO PROBLEMS SET

1. According to the law of conservation of mass, the mass of
    the product(s) must equal the mass of the reactant(s).
    Therefore, if 18.0 g Al reacts completely with 38.0 g F_2,
    56.0 g AlF_3 are formed.

2. H_2, O_2, N_2, F_2, Cl_2, Br_2 and I_2 are written as diatomic
    molecules when they appear as free elements in equations.

3. a) $N_2 + 3H_2 \longrightarrow 2NH_3$
    b) $2C + O_2 \longrightarrow 2CO$
    c) $2CO + O_2 \longrightarrow 2CO_2$
    d) $C_2H_4 + 3 O_2 \longrightarrow 2CO_2 + 2H_2O$
    e) $2C_2H_6 + 7 O_2 \longrightarrow 4CO_2 + 6H_2O$
    f) $2H_2O_2 \longrightarrow 2H_2O + O_2$
    g) $Na_2CO_3 + 2HCl \longrightarrow 2NaCl + H_2O + CO_2$
    h) $3Mg(NO_3)_2 + 2H_3PO_4 \longrightarrow Mg_3(PO_4)_2 + 6HNO_3$
    i) $FeCl_3 + 3KOH \longrightarrow 3KCl + Fe(OH)_3$
    j) $2Al(ClO_3)_3 \longrightarrow 2AlCl_3 + 9 O_2$

4. $\dfrac{2 \text{ moles } C_6H_{14}}{19 \text{ moles } O_2}$, $\dfrac{19 \text{ moles } O_2}{2 \text{ moles } C_6H_{14}}$

    $\dfrac{2 \text{ moles } C_6H_{14}}{12 \text{ moles } CO_2}$, $\dfrac{12 \text{ moles } CO_2}{2 \text{ moles } C_6H_{14}}$

    $\dfrac{2 \text{ moles } C_6H_{14}}{14 \text{ moles } H_2O}$, $\dfrac{14 \text{ moles } H_2O}{2 \text{ moles } C_6H_{14}}$

    $\dfrac{19 \text{ moles } O_2}{12 \text{ moles } CO_2}$, $\dfrac{12 \text{ moles } CO_2}{19 \text{ moles } O_2}$

    $\dfrac{19 \text{ moles } O_2}{14 \text{ moles } H_2O}$, $\dfrac{14 \text{ moles } H_2O}{19 \text{ moles } O_2}$

    $\dfrac{12 \text{ moles } CO_2}{14 \text{ moles } H_2O}$, $\dfrac{14 \text{ moles } H_2O}{12 \text{ moles } CO_2}$

5. a) 1.25 ~~moles CO~~ $\times \left(\dfrac{2 \text{ moles } CO_2}{2 \text{ ~~moles CO~~}} \right) = 1.25$ moles CO_2 (calculator and correct answer)

b) 1.25 ~~moles Ca(HCO$_3$)~~$_2$ $\times \left(\dfrac{2 \text{ moles } CO_2}{1 \text{ ~~mole Ca(HCO$_3$)~~}_2} \right)$

$= 2.5$ moles CO_2 (calculator answer)

$= 2.50$ moles CO_2 (correct answer)

c) 1.25 ~~moles C$_6$H$_{14}$~~ $\times \left(\dfrac{12 \text{ moles } CO_2}{2 \text{ ~~moles C$_6$H$_{14}$~~}} \right)$

$= 7.5$ moles CO_2 (calculator answer)

$= 7.50$ moles CO_2 (correct answer)

6. a) 5.48 ~~g C$_{25}$H$_{52}$~~ $\times \left(\dfrac{1 \text{ ~~mole C$_{25}$H$_{52}$~~}}{353 \text{ ~~g C$_{25}$H$_{52}$~~}} \right) \times \left(\dfrac{25 \text{ ~~moles CO$_2$~~}}{1 \text{ ~~mole C$_{25}$H$_{52}$~~}} \right)$

$\times \left(\dfrac{44.0 \text{ g } CO_2}{1 \text{ ~~mole CO$_2$~~}} \right) = 17.076487$ g CO_2 (calculator answer)

$= 17.1$ g CO_2 (correct answer)

b) 15.7 ~~g O$_2$~~ $\times \left(\dfrac{1 \text{ ~~mole O$_2$~~}}{32.0 \text{ ~~g O$_2$~~}} \right) \times \left(\dfrac{1 \text{ ~~mole C$_{25}$H$_{52}$~~}}{38 \text{ ~~moles O$_2$~~}} \right) \times \left(\dfrac{353 \text{ g } C_{25}H_{52}}{1 \text{ ~~mole C$_{25}$H$_{52}$~~}} \right)$

$= 4.557648$ g $C_{25}H_{52}$ (calculator answer)

$= 4.56$ g $C_{25}H_{52}$ (correct answer)

c) 5.48 ~~moles O$_2$~~ $\times \left(\dfrac{25 \text{ ~~moles CO$_2$~~}}{38 \text{ ~~moles O$_2$~~}} \right) \times \left(\dfrac{44.0 \text{ g } CO_2}{1 \text{ ~~mole CO$_2$~~}} \right)$

$= 158.63158$ g CO_2 (calculator answer)

$= 159$ g CO_2 (correct answer)

d) 1.57 ~~g C$_{25}$H$_{52}$~~ $\times \left(\dfrac{1 \text{ ~~mole C$_{25}$H$_{52}$~~}}{353 \text{ ~~g C$_{25}$H$_{52}$~~}} \right) \times \left(\dfrac{26 \text{ moles } H_2O}{1 \text{ ~~mole C$_{25}$H$_{52}$~~}} \right)$

$\times \left(\dfrac{6.02 \times 10^{23} \text{ molecules } H_2O}{1 \text{ ~~mole H$_2$O~~}} \right)$

$= 6.9613711 \times 10^{22}$ molecules H_2 (calculator answer)

$= 6.96 \times 10^{22}$ molecules H_2 (correct answer)

Calculations Involving Chemical Equations **157**

e) 6.02×10^{20} ~~molecules CO_2~~ $\times \left(\dfrac{1 \text{ mole } CO_2}{6.02 \times 10^{23} \text{ molecules } CO_2} \right)$

$\times \left(\dfrac{1 \text{ mole } C_{25}H_{52}}{25 \text{ moles } CO_2} \right) \times \left(\dfrac{353 \text{ g } C_{25}H_{52}}{1 \text{ mole } C_{25}H_{52}} \right)$

$= 1.412 \times 10^{-2}$ g $C_{25}H_{52}$ (calculator answer)

$= 1.41 \times 10^{-2}$ g $C_{25}H_{52}$ (correct answer)

7. a) 5.86 ~~g Cu~~ $\times \left(\dfrac{1 \text{ mole } Cu}{63.5 \text{ g } Cu} \right) \times \left(\dfrac{2 \text{ moles } H_2O}{1 \text{ mole } Cu} \right)$

$= 0.18456693$ moles H_2O (calculator answer)

$= 0.185$ moles H_2O (correct answer)

23.1 ~~g HNO_3~~ $\times \left(\dfrac{1 \text{ mole } HNO_3}{63.0 \text{ g } HNO_3} \right) \times \left(\dfrac{2 \text{ moles } H_2O}{4 \text{ moles } HNO_3} \right)$

$= 0.18333333$ moles H_2O (calculator answer)

$= 0.183$ moles H_2O (correct answer)

Therefore, HNO_3 is the limiting reactant and the amount of HNO_3 is used in all of the following calculations.

b) 23.1 ~~g HNO_3~~ $\times \left(\dfrac{1 \text{ mole } HNO_3}{63.0 \text{ g } HNO_3} \right) \times \left(\dfrac{2 \text{ moles } H_2O}{4 \text{ moles } HNO_3} \right) \times \left(\dfrac{18.0 \text{ g } H_2O}{1 \text{ mole } H_2O} \right)$

$= 3.3$ g H_2O (calculator answer)

$= 3.30$ g H_2O (correct answer)

c) 23.1 ~~g HNO_3~~ $\times \left(\dfrac{1 \text{ mole } HNO_3}{63.0 \text{ g } HNO_3} \right) \times \left(\dfrac{2 \text{ moles } NO_2}{4 \text{ moles } HNO_3} \right)$

$\times \left(\dfrac{6.02 \times 10^{23} \text{ molecules } NO_2}{1 \text{ mole } NO_2} \right)$

$= 1.1036667 \times 10^{23}$ molecules NO_2 (correct answer)

$= 1.10 \times 10^{23}$ molecules NO_2 (correct answer)

d) Cu is excess reactant.

$$23.1 \text{ g HNO}_3 \ \times \left(\frac{1 \text{ mole HNO}_3}{63.0 \text{ g HNO}_3}\right) \times \left(\frac{1 \text{ mole Cu}}{4 \text{ moles HNO}_3}\right) \times \left(\frac{63.5 \text{ g Cu}}{1 \text{ mole Cu}}\right)$$

= 5.8208333 g Cu reacts (calculator answer)
= 5.82 g Cu reacts (correct answer)

excess Cu is: 5.86 g Cu (starting amount)
                −5.82 g Cu (reacted)
                _____

= .04 g Cu = excess (calculator and
                     correct answer)

8. a) $12.5 \text{ g Fe(NO}_3)_3 \ \times \left(\frac{1 \text{ mole Fe(NO}_3)_3}{242 \text{ g Fe(NO}_3)_3}\right) \times \left(\frac{1 \text{ mole Fe(OH)}_3}{1 \text{ mole Fe(NO}_3)_3}\right)$

$\times \left(\frac{107 \text{ g Fe(OH)}_3}{1 \text{ mole Fe(OH)}_3}\right)$ = 5.5268595 g Fe(OH)$_3$
                                    (calculator answer)
                              = 5.53 g Fe(OH)$_3$
                                    (correct answer)

b) percent yield = $\dfrac{\text{actual yield}}{\text{theoretical yield}}$ x 100

= $\dfrac{5.26 \text{ g}}{5.53 \text{ g}}$ x 100 = 95.117541%
                          (calculator answer)
                    = 95.1% (correct answer)

Optional

9. $86.3 \text{ g mixture} \ \times \left(\frac{15.9 \text{ g Ag}_2\text{O}}{100 \text{ g mixture}}\right)$ = 13.7217 g Ag$_2$O
                                    (calculator answer)
                              = 13.7 g Ag$_2$O
                                    (correct answer)

and

$86.3 \text{ g mixture} \ \times \left(\frac{84.1 \text{ g HgO}}{100 \text{ g mixture}}\right)$ = 72.5783 g HgO
                                    (calculator answer)
                              = 72.6 g HgO
                                    (correct answer)

From the reaction: $2\text{Ag}_2\text{O} \longrightarrow 4\text{Ag} + \text{O}_2$

$13.7 \text{ g Ag}_2\text{O} \ \times \left(\frac{1 \text{ mole Ag}_2\text{O}}{232 \text{ g Ag}_2\text{O}}\right) \times \left(\frac{1 \text{ mole O}_2}{2 \text{ mole Ag}_2\text{O}}\right) \times \left(\frac{32.0 \text{ g O}_2}{1 \text{ mole O}_2}\right)$

= 0.94482759 g O$_2$ (calculator answer)
= 0.945 g O$_2$ (correct answer)

From the reaction: $2HgO \longrightarrow 2Hg + O_2$

$$72.6 \; g \; HgO \; \times \left(\frac{1 \; mole \; HgO}{217 \; g \; HgO}\right) \times \left(\frac{1 \; mole \; O_2}{2 \; mole \; HgO}\right) \times \left(\frac{32.0 \; g \; O_2}{1 \; mole \; O_2}\right)$$

$$= 5.3529954 \; g \; O_2 \; \text{(calculator answer)}$$
$$= 5.35 \; g \; O_2 \; \text{(correct answer)}$$

The total weight of O_2 produced is:

$$0.945 \; g \; O_2 \; \text{(from } Ag_2O\text{)}$$
$$+5.35 \;\; g \; O_2 \; \text{(from HgO)}$$
$$\overline{}$$
$$= 6.295 \; g \; O_2 \; \text{(calculator answer)}$$
$$= 6.30 \; g \; O_2 \; \text{(correct answer)}$$

10. $17.0 \; g \; NaNO_3 \; \times \left(\frac{1 \; mole \; NaNO_3}{85.0 \; g \; NaNO_3}\right) \times \left(\frac{1 \; mole \; O_2}{2 \; moles \; NaNO_3}\right) \times \left(\frac{2 \; moles \; CO_2}{1 \; mole \; O_2}\right)$

$$\times \left(\frac{44.0 \; g \; CO_2}{1 \; mole \; CO_2}\right) = 8.8 \; g \; CO_2 \; \text{(calculator answer)}$$
$$= 8.80 \; g \; CO_2 \; \text{(correct answer)}$$

ANSWERS TO MULTIPLE CHOICE EXERCISES

1.	c	12.	a
2.	d	13.	b
3.	b	14.	b
4.	c	15.	a
5.	b	16.	d
6.	b	17.	a
7.	d	18.	a
8.	b	19.	b
9.	d	20.	b
10.	c	21.	c
11.	d	22.	c

CHAPTER 9
The Electronic Structure
of Atoms

REVIEW OF CHAPTER OBJECTIVES

1. Explain what is meant by quantized energy (section 9.1).

   The energy of an electron is a quantized property, which means that it may have only certain allowable values.

2. Understand how the terms shell, subshell and orbital are used in describing electron arrangements about the nucleus, and know the mathematical interrelationships between them (sections 9.2-9.4).

   Electron Shells

   Higher energy electrons will be found farther from the nucleus than lower energy electrons. Electrons are grouped into shells, or main energy levels, based on considerations of their energy and distance from the nucleus. A SHELL contains electrons that have approximately the same energy, and that spend most of their time approximately the same distance from the nucleus.

   Electron shells are identified by either of two systems. The older system uses capital letters of the alphabet starting with K (K, L, M, N, etc.) and the modern system uses numbers starting with 1 (1, 2, 3, 4, etc.). In either case, the first letter or number indicates the lowest energy shell, the one that is closest to the nucleus.

Shell designation	older method	newer method
lower energy - closer to nucleus	K	1
	L	2
	M	3
	N	4
	O	5
	P	6
higher energy - farther from nucleus	Q	7

No known element has electrons in shells beyond Q (7th shell).

The maximum number of electrons found in an electron shell increases as the energy increases. The higher the energy of the shell the greater the space occupied by the electrons, and the larger the maximum number of electrons in that shell.

The maximum number of electrons in a shell $= 2n^2$
where n = the number of the shell

Shell	Letter Designation	Number designation (n)	Maximum number of electrons $= 2n^2$
first	K	1	2
second	L	2	8
third	M	3	18
fourth	N	4	32
fifth	O	5	(50) beyond the
sixth	P	6	(72) fourth shell, no
seventh	Q	7	(98) element yet dis-

beyond the fourth shell, no element yet discovered has attained the maximum number of electrons shown.

Electron Subshells

All electrons in a shell do not have the same energies. The energies are slightly different, due to the existence of subshells, or sub-energy levels. A SUBSHELL contains electrons which all have identical energies.

A given shell contains as many subshells as its shell number designation. Each successive shell has one more subshell than the previous shell.

The subshells are identified by a number and a letter (lower case). The number indicates the shell to which the subshell belongs, and the lower case letter is either s, p, d or f. The s subshell is the lowest energy sublevel in each shell, followed by p (the next lowest), etc.

Shell 1 has only 1 subshell, 1s
Shell 2 has 2 subshells, 2s and 2p
Shell 3 has 3 subshells, 3s and 3p and 3d
Shell 4 has 4 subshells, 4s and 4p and 4d and 4f

None of the known (106) elements have electrons in subshells beyond the first 4 (s, p, d and f).

An s subshell, in any shell, holds no more than 2 electrons.

A p subshell, in any shell, holds no more than 6 electrons.

A d subshell, in any shell, holds no more than 10 electrons.

An f subshell, in any shell, holds no more than 14 electrons.

Electron Orbitals

An electron ORBITAL is a region of space around a nucleus where an electron with a given energy is most likely to be found.

An s subshell consists of 1 orbital
A p subshell consists of 3 orbitals
A d subshell consists of 5 orbitals
An f subshell consists of 7 orbitals

Different subshells have different numbers of orbitals (from 1 orbital in an s subshell to 7 orbitals in an f subshell). However, an orbital (in an s, or p, or d, or f subshell) holds a maximum of 2 electrons.

All electrons in a subshell have the same energy. The energy of an electron is specified by stating its shell and subshell location.

Orbitals have a definite shape and size, related to the subshell where they are found. Figure 9-2 on page 206 of the textbook shows various orbital shapes. Orbitals of the same type, found in different shells, have the same shape but differ in size. For example, a 3s orbital is larger (in volume) than a 2s orbital, which in turn is larger than a 1s orbital.

Example: Determine the following for the third electron shell (third main electron energy level) of an atom:

    a) the number of subshells it contains
    b) the designation used to describe each subshell
    c) the number of orbitals in each subshell
    d) the maximum number of electrons that could be contained in each subshell
    e) the maximum number of electrons that could be contained in the shell

Solution:

    a) The number of subshells in a shell equals the shell number; therefore, the third shell contains 3 subshells.

    b) The subshell designations, from lowest to highest in energy, in the third shell, are:

        3s, 3p, 3d

    c) The number of orbitals in each subshell is independent of the shell number.

        Any:

        s subshell contains 1 orbital, so the 3s subshell has 1 orbital
        p subshell contains 3 orbitals, so the 3p subshell has 3 orbitals
        d subshell contains 5 orbitals, so the 3d subshell has 5 orbitals

    d) The maximum number of electrons in any orbital is 2. Therefore:

        a 3s subshell has 1 orbital and 2 electrons
        a 3p subshell has 3 orbitals and 6 electrons
        a 3d subshell has 5 orbitals and 10 electrons

    e) The maximum number of electrons in a shell $= 2n^2$

        since $n = 3$, $2n^2 = 2 \times 3^2 = 2 \times 9 = 18$

        Alternately, adding up the electrons shown in part d) above gives:
        $2 + 6 + 10 = 18$ electrons

3. <u>Write out the electron configurations for any element</u>
<u>using the Aufbau principle</u> (section 9.5).

An <u>electron configuration</u> states how many electrons an atom has in each of its subshells. Number-letter combinations are used to list subshells which contain electrons, in order of increasing energy. A superscript following each subshell designation indicates the number of electrons in that subshell.

As an example, the electron configuration for nitrogen is:

$$1s^2 \; 2s^2 \; 2p^3$$

This notation indicates the presence of 2 electrons in the 1s subshell, 2 in the 2s subshell and 3 electrons in the 2p subshell. The Aufbau Principle is used to find an atom's electron configuration. The Aufbau principle states that electrons occupy the lowest energy sublevel available. Orbitals are filled in order of increasing energy. The order in which subshells fill is given on page 210 of the textbook in terms of two different subshell diagrams (Figures 9-5 and 9-6). Most often, the second of the two diagrams, the Aufbau diagram, is used to determine the order of filling.

To construct an Aufbau diagram, one arranges the subshells, evenly spaced, with those in higher energy shells placed below each other, as follows:

```
1s
2s    2p
3s    3p    3d
4s    4p    4d    4f
5s    5p    5d    5f
6s    6p    6d    6f
7s    7p    7d    7f
```

Then diagonal parallel lines are drawn, from upper right to lower left, through the subshells in the order in which they fill with electrons. You know the order of filling of the first 4 subshells:

1s, then 2s, then 2p, then 3s

Therefore, you draw the first three parallel lines to indicate this order, with arrows on the lower left end.

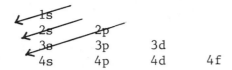

Then draw the other parallel lines, in a similar way:

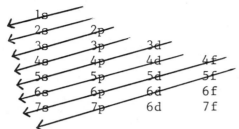

This diagram shows that the order of filling is:

1s 2s 2p 3s 3p 4s 3d 4p 5s etc.

To use this diagram, you start at the tail (upper right end) of the top arrow and follow it down to its head before moving to the tail of the next arrow.

Example: Write out the electron configurations for:

a) P (Z = 15)
b) Se (Z = 34)
c) In (Z = 49)

Solution:

a) P has 15 electrons.

$$1s^2 2s^2 2p^6 3s^2 3p^3$$

b) Se has 34 electrons.

$$1s^2 2s^2 2p^6 3s^2 3p^6 4s^2 3d^{10} 4p^4$$

c) In has 49 electrons.

$$1s^2 2s^2 2p^6 3s^2 3p^6 4s^2 3d^{10} 4p^6 5s^2 4d^{10} 5p^1$$

Note that, besides filling the subshells according to the Aufbau principle, the total number of electrons shown must equal the atomic number of the element (no more or less).

a) 2+2+6+2+3 = 15

b) 2+2+6+2+6+2+10+4 = 34

c) 2+2+6+2+6+2+10+6+2+10+1 = 49

For a few elements, located in the middle of the periodic
table, the actual electron distribution differs slightly
from the prediction obtained according to the Aufbau
principle, due to very small differences in energy
between some subshells. However, we can ignore these
exceptions in this beginning course.

Abbreviated electron configurations, which give only the
number of electrons per shell, are sometimes of interest.
The abbreviated electron configurations for P, Se and In
are:

P = 2, 8, 5

Se = 2, 8, 18, 6

In = 2, 8, 18, 18, 3

4. Optional. List the names, symbols and functions of the
four quantum numbers and assign a physical meaning to
each (section 9.6).

The previous sections describe electron arrangements
semiquantitatively. A more quantitative approach
describes electron configurations using quantum numbers,
numbers which describe the allowed energy states of
electrons in orbitals.

A given electron is described using a set of four
quantum numbers.

The principal quantum number is a positive integer:

n = 1, 2, 3, 4, etc.

Seven is the highest value for n for electrons in
currently known atoms.

The principal quantum number indicates the shell in which
the electron is found (and it is the same as the shell
number previously discussed.) The larger the principal
quantum number, the larger the orbital in which the
electron described is located.

The secondary quantum number, ℓ , has allowable values
determined by n. ℓ can be any whole number between 0
and n-1:

$$\ell = 0, 1, 2, 3, \ldots (n - 1)$$

Value of n	Values of ℓ
1	0
2	0, 1
3	0, 1, 2

The secondary quantum number divides shells into subshells.

Shell number 3 (n = 3) contains 3 subshells:

the 3s subshell ($\ell = 0$)
the 3p subshell ($\ell = 1$), and
the 3d subshell ($\ell = 2$)

Note the match-up between subshell letter designations and secondary quantum number values.

letter subshell designation	Value of ℓ (secondary quantum number)
s	0
p	1
d	2
f	3

All electrons in a given subshell must have the same n and ℓ values. In the 4f subshell, all electrons have the values:

$$n = 4 \text{ and } \ell = 3$$

The secondary quantum number specifies the shape of the orbital in which an electron is found.

The <u>magnetic quantum number</u>, m, is the third quantum number. The restriction on m is that its values must be integers, that range from $-\ell$ through zero to $+\ell$.

$$m = 0, \pm 1, \pm 2, \ldots \pm \ell$$

For example, when $\ell = 3$ (an f subshell), m can have 7 possible values:

$$-3, -2, -1, 0, +1, +2, +3$$

The magnetic quantum number splits subshells into individual orbitals, and specifies the arrangement in space of an orbital within a subshell.

An orbital holds a maximum of 2 electrons, both having the same n, ℓ, and m values.

The spin quantum number, s, is the fourth quantum number, and can have only 2 values, +1/2 or -1/2. The spin quantum number distinguishes the 2 electrons in a given orbital.

Each of the 4 quantum numbers conveys specific information about an electron:

Principal quantum number, n, locates the electron's shell and determines the size of the orbital where the electron is found.

Secondary quantum number, ℓ, locates the electron's subshell and specifies the shape of the orbital where the electron is found.

Magnetic quantum number, m, locates the specific orientation in space of the orbital where the electron is found.

Spin quantum number, s, distinguishes between 2 electrons in the same orbital, and specifies spin (clockwise or counterclockwise) of an electron.

5. Optional. Draw orbital diagrams consistent with the Pauli exclusion principle and Hund's rule (section 9.7).

The Aufbau principle specifies the arrangement of electrons in shells and subshells. In addition to the Aufbau principle, 2 other principles are needed to specify orbital occupancy for electrons. These other 2 principles are:

The Pauli exclusion principle, which states that no two electrons in an atom may have the same set of 4 quantum numbers. Since two electrons in an orbital have the same values of n, ℓ and m, they must differ in spin. Two electrons in an orbital, which differ only in spin, are said to be "paired."

Hund's rule, the principle of maximum multiplicity, states that electrons fill equal energy orbitals one electron at a time. This means that one electron goes into each one of the equal energy orbitals first, before a second electron goes into any one of these equal energy orbitals.

Orbital diagrams show electron occupancy of each

orbital in an atom, with a circle representing an
orbital and a single-barbed arrow denoting an electron.

H ① $(1s^1)$

He ⑴ $(1s^2)$

Li ⑴ ① $(1s^2 2s^1)$

Be ⑴ ⑴ $(1s^2 2s^2)$

Starting with B, atomic number 5, an electron goes into
one of several equal energy (2p) orbitals (all
equivalent orbitals are shown in the diagram even if not
all of them contain an electron).

B ⑴ ⑴ ① ○○ $(1s^2 2s^2 2p^1)$

Starting with C, we see Hund's rule apply.

C ⑴ ⑴ ①① ○ $(1s^2 2s^2 2p^2)$

The second electron in the 2p subshell goes into the
second 2p orbital, unpaired, rather than pairing up in
the first 2p orbital that already has one electron.

Example: Write orbital diagrams for P, Ca and V.

Solution:

For P:

P has the electron configuration:

$1s^2 2s^2 2p^6 3s^2 3p^3$

The orbital diagram is:

P ⑴ ⑴ ⑴⑴⑴ ⑴ ①①①
   1s   2s   2p   3s   3p

For Ca:

Ca has the electron configuration:

$1s^2 2s^2 2p^6 3s^2 3p^6 4s^2$

The orbital diagram is:

Ca:

For V:

V has the electron configuration:

$1s^2 2s^2 2p^6 3s^2 3p^6 4s^2 3d^3$

The orbital diagram is:

Atoms are said to be paramagnetic if they contain orbitals with one or more unpaired electrons, as in P and V (above).

Atoms are diamagnetic if all electrons are paired in orbitals, as in Ca (above).

6. <u>Optional. Use the (n + ℓ) rule to predict the order of filling of electron subshells</u> (section 9.8).

In order to determine electron configurations or orbital diagrams, the relative energies of orbitals and shells must be known. The Aufbau diagram is a useful mnemonic device. However, a second method using quantum numbers is also available. This method, called the (n + ℓ) rule, uses the sum of the values of n (the principal quantum number) and ℓ (the secondary quantum number) to predict relative energies. The two parts of the (n + ℓ) rule are that:

1) The orbital with the lowest sum of n + ℓ will be occupied by electrons first.

2) If two orbitals have the same n + ℓ sum, the one with the lower n value will be occupied by electrons first.

Example: Use the (n + ℓ) rule to determine whether:

1) the 4d or the 5s subshell is the first to be occupied by electrons.

2) the 4f or the 5d subshell is the first to be occupied by electrons.

Solution:

1)

Orbital	n value	ℓ value	$(n + \ell)$ sum
4d	4	2	6
5s	5	0	5

The $(n + \ell)$ value for 5s is lower than the one for 4d, so the 5s subshell has a lower energy and it is occupied by electrons first.

2)

Orbital	n value	ℓ value	$(n + \ell)$ sum
4f	4	3	7
5d	5	2	7

The 4f and the 5d subshells both have an $(n + \ell)$ sum of 7. So the one with the lower n value, the 4f, is of lower energy, and thus the 4f subshell is occupied by electrons first.

7. Understand the relationship between the periodic law and electron configurations (section 9.9).

According to the periodic law, when elements are arranged in order of increasing atomic number, their properties repeat in a regular manner. Those elements with similar chemical properties are found placed one under another in vertical columns (groups) of the periodic table. Groups of elements have similar chemical properties because these elements have "similar" electron configurations.

As an example, we consider the elements of group IIA, which are known for having similar properties.

Be $1s^2$ $2s^2$

Mg $1s^2$ $2s^2$ $2p^6$ $3s^2$

Ca $1s^2$ $2s^2$ $2p^6$ $3s^2$ $3p^6$ $4s^2$

Sr $1s^2$ $2s^2$ $2p^6$ $3s^2$ $3p^6$ $4s^2$ $3d^{10}$ $4p^6$ $5s^2$

All have a filled outer s orbital; the similarities in properties are due to the similar outer shell electron configuration.

8. Use the periodic table to determine the extent of filling of the subshell containing the distinguishing electron and the shell in which this subshell is located for any

The concept of distinguishing electrons is the key to obtaining electron configuration information from the periodic table. The distinguishing electron for an element is the last electron added to its electron configuration using the Aufbau principle. This last electron added is the one that causes an element's electron configuration to differ from that of the element immediately preceding it in the periodic table.

The subshell containing the element's distinguishing electron, and the extent of filling of that subshell, can be determined from the location of the element in the periodic table, as illustrated in Figures 9.8 and 9.9 in the textbook (pages 224 and 225).

Example: For the elements "A", "B", "C", "D" and "E" in the following periodic table, indicate:

a) whether the distinguishing electron is in the s, p, d or f subshell.

b) the extent of filling of the subshell containing the distinguishing electron.

c) the shell in which the distinguishing electron is found.

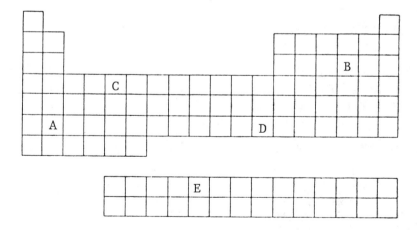

Solution:

"A" is in the s area, so its distinguishing electron is in the s subshell; there are 2 electrons in the s subshell, because "A" is in column 2 of the s area; and the distinguishing electron is in the 6th shell since the element is in period 6.

"B" is in the p area and has its distinguishing electron in the p subshell. There are 4 electrons in this p subshell because "B" is in column 4 of the p area, and the distinguishing electron is in the third shell since the element is in period 3.

"C" has the distinguishing electron in the d subshell. There are 3 electrons in this d subshell and the distinguishing electron is in the third shell (n = 3) because in the d area, the period number minus one equals the shell number.

"D" has the distinguishing electron in the d subshell. There are 10 electrons in this d subshell and the distinguishing electron is in the fifth shell (period number minus one).

"E" has the distinguishing electron in the f subshell. There are 5 electrons in this f subshell and the distinguishing electron is in the fourth shell (n = 4) because in the f area, the period number minus two equals the shell number.

9. Use the periodic table as a guide in writing electron configurations (section 9.10).

To write complete electron configurations, the order of filling of the various electron subshells must be known. We determine the order of filling of subshells from the Aufbau diagram or the (n + ℓ) rule; but it can also be determined directly by "reading" the information shown on the periodic table. The latter technique involves following a path of increasing atomic number through the periodic table noting the various subshells as they are encountered. The results of this procedure are summarized in Figure 9-10 (page 226) and Table 9-11 (page 227) in the textbook. Another example of this procedure is as follows:

Example: Use the periodic table as your sole guide to write the complete electron configuration of Os (Z = 76).

Solution:

Working through the periodic table, starting with atomic number 1 and adding one electron at a time in order of increasing atomic number, we find there are:

2 electrons in the 1s subshell (atomic numbers 1-2)
2 electrons in the 2s subshell (atomic numbers 3-4)
6 electrons in the 2p subshell (atomic numbers 5-10)
2 electrons in the 3s subshell (atomic numbers 11-12)
6 electrons in the 3p subshell (atomic numbers 13-18)
2 electrons in the 4s subshell (atomic numbers 19-20)
10 electrons in the 3d subshell (atomic numbers 21-30)
6 electrons in the 4p subshell (atomic numbers 31-36)
2 electrons in the 5s subshell (atomic numbers 37-38)
10 electrons in the 4d subshell (atomic numbers 39-48)
6 electrons in the 5p subshell (atomic numbers 49-54)
2 electrons in the 6s subshell (atomic numbers 55-56)
14 electrons in the 4f subshell (atomic numbers 58-71)
6 electrons in the 5d subshell (atomic numbers 57 and 72-76)

Total = 76 electrons

Os: $1s^2 2s^2 2p^6 3s^2 3p^6 4s^2 3d^{10} 4p^6 5s^2 4d^{10} 5p^6 6s^2 4f^{14} 5d^6$

10. <u>Classify a given element as a rare gas, representative element, transition element or rare earth element</u> (section 9.11).

Elements can be classified in several ways. One way, as metal or nonmetal, divides elements into 2 categories, depending on whether they appear on the periodic table to the left of (metals) or to the right of (nonmetals) the staircase-shaped line (or pattern) that runs between B and Al, Al and Si, Si and Ge, etc.

The second way is based on electron configuration, and describes elements as:

Representative elements -- all elements of the s and p areas of the periodic table with the exception of the last column of the p area (the rare gases).

Rare gas elements -- the elements in the last column of the p area of the periodic table.

Transition elements -- all of the elements in the d area of the periodic table.

Rare earth elements -- all of the elements in the f area

of the periodic table.

Example: From the position on the periodic table, classify the following elements as representative elements, rare gases, transition elements or rare earth elements:

Ba (Z = 56), La (Z = 57), Er (Z = 68), Re (Z = 75), Au (Z = 79), Pb (Z = 82), Rn (Z = 86) and Fr (Z = 87).

Solution:

In s and p areas, except for the last column on the extreme right of the periodic table, the elements are representative elements.

Ba (Z = 56) and Fr (Z = 87) in the s area, and Pb (Z = 82) in the p area are representative elements.

Rn (Z = 86) is a rare gas, being in the right-hand column of the p area.

Re (Z = 75) and Au (Z = 79), in the d area, are transition elements.

La (Z = 57) and Er (Z = 68), in the f area, are rare earth elements.

PROBLEM SET

1. Determine the following for the fifth electron shell (the fifth main energy level) of an atom:

    a) the number of subshells it contains
    b) the designation used to describe each of the first four subshells
    c) the number of orbitals in each of the first four subshells
    d) the maximum number of electrons that can occupy this fifth shell
    e) the maximum number of electrons that can occupy each of the first four subshells

2. Fill in the numerical value that correctly completes each statement:

    a) the maximum number of electrons in the second energy level is _____ .

b) the maximum number of electrons in the sixth energy
   level is _____.
c) a 4f subshell holds a maximum of _____ electrons.
d) a 3d orbital holds a maximum of _____ electrons.
e) a 2d subshell holds a maximum of _____ electrons.
f) the fifth shell contains _____ subshells, _____
   orbitals and a maximum of _____ electrons.

3. Using the Aufbau principle, write electron configurations
   for:

   a) Cl (Z = 17)
   b) Sn (Z = 50)
   c) Ho (Z = 67)

4. For each of the following elements, tell:

   a) whether the distinguishing electron is in the s, p, d
      or f subshell
   b) the extent of filling of the subshell containing the
      distinguishing electron
   c) the shell in which the distinguishing electron is
      found

      1) Fe (Z = 26)
      2) Xe (Z = 54)
      3) Pr (Z = 59)
      4) Ra (Z = 88)

5. Identify the element that has each of the following
   electron configurations:

   a) $1s^2 2s^2 2p^3$
   b) $1s^2 2s^2 2p^6 3s^2 3p^6 4s^2 3d^{10}$
   c) $1s^2 2s^2 2p^6 3s^2 3p^6 4s^2 3d^{10} 4p^6 5s^2 4d^{10} 5p^6$
   d) $1s^2 2s^2 2p^6 3s^2 3p^6 4s^2 3d^{10} 4p^6 5s^2 4d^{10} 5p^6 6s^2 5d^1 4f^7$

6. Indicate the position in the periodic table (by giving
   the symbol of the element) where each of the following
   occurs:

   a) the 3s subshell becomes completely filled
   b) the 4p subshell begins filling
   c) the 5d subshell begins filling
   d) the 2p subshell becomes half-filled
   e) the fourth shell begins filling
   f) the fourth shell becomes completely filled
   g) the second shell becomes half-filled

7. Using only the periodic table write the complete
    electron configuration of:

    a) Cd (Z = 48)
    b) La (Z = 57)

8. Indicate whether each of the following elements is a
    representative element, rare gas, transition element or
    rare earth element:

    a) $_8O$ d) $_{47}Ag$

    b) $_{10}Ne$ e) $_{92}U$

    c) $_{19}K$ f) $_2He$

Optional Problems

9. List all 4 quantum numbers for each electron in an atom
    of:

    a) N
    b) Mg

10. Draw orbital diagrams, consistent with Pauli's exclusion
    principle and Hund's rule, for an atom of:

    a) S
    b) K
    c) N

11. Use the $(n + \ell)$ rule to predict which subshell in each
    of the following pairs of subshells is occupied first by
    electrons:

    a) 5s or 4d
    b) 5s or 4p
    c) 5d or 4f

MULTIPLE CHOICE EXERCISES

1. The space around a nucleus in which electrons move can
    be classified into several subspaces. Which of the
    following is the most complex of these classifications?

    a) orbital
    b) shell
    c) subshell
    d) apartment

2. The formula $2n^2$ gives the maximum number of electrons which may occupy:

   a) an orbital
   b) a shell
   c) a subshell
   d) an apartment

3. The term <u>energy sublevel</u> is closely associated with the term:

   a) orbital
   b) shell
   c) subshell
   d) supershell

4. Both a number and a letter are used in designating a subshell. The letter:

   a) indicates the shell to which the subshell belongs
   b) may be s, p, f or r
   c) gives information about the maximum number of electrons the subshell may hold
   d) is always placed before the number

5. Which of the following statements about an f subshell is correct?

   a) it contains 5 orbitals
   b) it may contain a maximum of 14 electrons
   c) it is only found in shells 5 or 6
   d) it determines the shape of an atom

6. The number of possible subshells in a shell is:

   a) equal to the shell number
   b) one less than the shell number
   c) twice the shell number
   d) equal to $2n^2$ where n is the shell number

7. Which series of subshells is arranged in order of increasing energy in an atom with many electrons?

   a) 3d, 4s, 4p, 5s
   b) 4s, 3d, 4p, 5s
   c) 4s, 5s, 4p, 3d
   d) 3d, 4p, 4s, 5s

8. Which of the following statements is consistent with the electronic configuration $1s^2 2s^2 2p^6 3s^2 3p^6$?

a) there are 6 electrons in the 3p orbital
b) there are 6 electrons in the 3p subshell
c) there are 6 electrons in the 3s orbital
d) there are 6 electrons in the 3p supershell

9. The correct electron configuration for phosphorus (element #15) is:

a) $1s^2 2s^2 2p^6 3s^2 3p^3$
b) $1s^2 2s^2 2p^6 3s^2 3p^5$
c) $1s^2 2s^2 2p^6 3s^2 3p^6$
d) $1s^2 2s^2 2p^6 3s^2 3p^6 4s^2 4p^3$

10. The correct electron configuration for iodine (element #53) is:

a) $1s^2 2s^2 2p^6 3s^2 3p^6 4s^2 4p^6 5s^2 5p^5$
b) $1s^2 2s^2 2p^6 3s^2 3p^6 3d^{10} 4s^2 4p^6 4d^{10} 4f^7$
c) $1s^2 2s^2 2p^6 3s^2 3p^6 4s^2 3d^{10} 4p^6 5s^2 4d^{10} 5p^5$
d) $1s^2 2s^2 2p^6 3s^2 3p^6 4s^2 3d^{10} 4p^6 4f^{14} 5p^3$

11. All of the following attempts to write electron configurations are incorrect except for one. The correct one is:

a) $1s^2 2s^2 2p^8 3s^2 3p^8$
b) $1s^2 2s^2 2p^6 3p^3$
c) $1s^2 2s^2 2p^3$
d) $1s^2 2s^2 2p^6 3s^2 3p^6 3d^{10}$

12. Electron configurations for elements falling the same group in the periodic table have what in common?

a) same number of electrons in all subshells
b) outermost electrons are always in an s subshell
c) same number of electrons in the outermost occupied subshell
d) all subshells are completely filled

13. Which of the following electron configurations corresponds to an element in the same group of the periodic table as the element whose electron configuration is $1s^2 2s^2 2p^5$?

    a) $1s^2 2s^2 2p^4$

    b) $1s^2 2s^2 2p^6$

    c) $1s^2 2s^2 2p^6 3s^2 3p^5$

    d) $1s^2 2s^2 2p^6 3s^1$

14. Which of the following elements has an electronic configuration ending in d^7?

    a) $_7 N$

    b) $_{27} Co$

    c) $_{47} Ag$

    d) $_{51} Sb$

15. The elements in group VA on the periodic table all have electron configurations ending in

    a) p^3

    b) p^5

    c) d^5

    d) s^2

16. Which of the following pairings is a correct pairing?

    a) $_{37} Rb$ -- p area of periodic table

    b) $_{75} Re$ -- f area of periodic table

    c) $_{13} Al$ -- s area of periodic table

    d) $_{80} Hg$ -- d area of periodic table

17. What common feature characterizes the electronic configuration of elements 21, 27, 40 and 43?

    a) they all have only one electron in an s subshell
    b) they all have a filled f subshell
    c) they all have a partially filled d subshell
    d) they all have a partially filled p subshell

18. Which of the following element-classification pairings is incorrect?

    a) $_{12}Mg$ -- representative element
    b) $_{17}Cl$ -- rare gas
    c) $_{47}Ag$ -- transition element
    d) $_{58}Ce$ -- rare earth element

Optional (sections 9.6 - 9.8 of text)

19. The rule that states no two electrons can have the same four quantum numbers is known as:

    a) Hund's rule
    b) Pauli exclusion principle
    c) Aufbau's law
    d) periodic law

20. Which of the following is a secondary quantum number?

    a) n = 3
    b) s = 1/2
    c) ℓ = 2
    d) m = 0

21. Which of the following is an "impossible" set of quantum numbers for an electron?

    a) 2, 1, -1, +1/2
    b) 4, 3, -2, +1/2
    c) 1, -1, 0, +1/2
    d) 3, 0, 0, -1/2

22. Which of the following atoms contains 2 unpaired electrons:

    a) $_7N$
    b) $_8O$
    c) $_9F$
    d) $_{10}Ne$

23. What are the values of n and ℓ for an electron in a 3p orbital?

    a) n = 3, ℓ = 0
    b) n = 3, ℓ = 1
    c) n = 1, ℓ = 1
    d) n = 1, ℓ = 3

182   The Electronic Structure of Atoms

24. What is the maximum number of electrons that an atom can have with the quantum number specification n = 3, ℓ = 2, m = 2?

    a) 2
    b) 3
    c) 6
    d) 18

PROBLEM SET SOLUTIONS

1. a) The number of subshells in a shell equals the shell number. Therefore, shell number 5 contains 5 subshells.

    b) The subshell designation (of the first 4 subshells) in the fifth shell is 5s, 5p, 5d and 5f. The fifth subshell has not been named, and no known atom has a shell containing electrons in more than 4 subshells.

    c) The number of orbitals in each of the first 4 subshells is:

    any s subshell contains 1 orbital
    any p subshell contains 3 orbitals
    any d subshell contains 5 orbitals
    any f subshell contains 7 orbitals

    d) The maximum number of electrons in a shell is $2n^2$. If n = 5, then $2n^2 = 2 \times 5^2 = 2 \times 25 = 50$

    e) The maximum number of electrons in each of the first 4 subshells (of the fifth energy level) is:

    an s subshell, with 1 orbital, has 2 electrons
    a p subshell has 3 orbitals and 6 electrons
    a d subshell has 5 orbitals and 10 electrons
    an f subshell has 7 orbitals and 14 electrons

2. a) 8
    b) 32
    c) 14
    d) 2 (any orbital has a maximum occupancy of 2 electrons)
    e) 0 (a 2d orbital doesn't exist)
    f) 5 subshells, 25 orbitals and 50 electrons

3. The order of filling shells and subshells, according to the Aufbau principle, is:

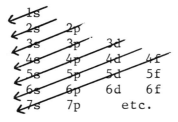

Therefore,

a) Cl has 17 electrons:

$$1s^2 2s^2 2p^6 3s^2 3p^5$$

b) Sn has 50 electrons:

$$1s^2 2s^2 2p^6 3s^2 3p^6 4s^2 3d^{10} 4p^6 5s^2 4d^{10} 5p^2$$

c) Ho has 67 electrons:

$$1s^2 2s^2 2p^6 3s^2 3p^6 4s^2 3d^{10} 4p^6 5s^2 4d^{10} 5p^6 6s^2 5d^1 4f^{10}$$

4. 1) Fe, with 26 electrons, has its distinguishing electron in the d subshell, which contains 6 electrons, and in the third shell (n = 3).

   2) Xe, with 54 electrons, has its distinguishing electron in the p subshell, which has 6 electrons, and in the fifth shell (n = 5).

   3) Pr, with 59 electrons, has its distinguishing electron in the f subshell, which has 2 electrons, and in the fourth shell (n = 4).

   4) Ra, with 88 electrons, has its distinguishing electron in the s subshell, which contains 2 electrons, and in the seventh shell (n = 7).

5. a) $_7 N$

   b) $_{30} Zn$

   c) $_{54} Xe$

   d) $_{64} Gd$

6. a) $_{12} Mg$

   b) $_{31} Ga$

c) $_{57}La$

d) $_7N$ (for a p subshell, half-filled is 3 electrons)

e) $_{19}K$ (the first electrons into the 4th shell are the 4s)

f) $_{71}Lu$ (the fourth shell contains 4s, 4p, 4d and 4f subshells; the 4f is the last of the four to fill.

g) $_6C$ (the second shell contains 8 electrons $(2s^2 2p^6)$. A $2s^2 2p^2$ configuration (4 electrons) would be half-filled.)

7. a) Cd, Z = 48, has 48 electrons.

   2 electrons in 1s (atomic number 1-2)
   2 electrons in 2s (atomic number 3-4)
   6 electrons in 2p (atomic number 5-10)
   2 electrons in 3s (atomic number 11-12)
   6 electrons in 3p (atomic number 13-18)
   2 electrons in 4s (atomic number 19-20)
   10 electrons in 3d (atomic number 21-30)
   6 electrons in 4p (atomic number 31-36)
   2 electrons in 5s (atomic number 37-38) and
   <u>10 electrons</u> in 4d (atomic number 39-48)

   Total = 48 electrons.

   Cd: $1s^2 2s^2 2p^6 3s^2 3p^6 4s^2 3d^{10} 4p^6 5s^2 4d^{10}$

   b) La, Z = 57, has 57 electrons. The first 48 as in Cd above, plus:

   6 electrons in 5p (atomic number 49-54)
   2 electrons in 6s (atomic number 55-56) and
   <u>1 electron in</u> 4d (atomic number 57)

   Total = 48 + 9 = 57 electrons.

   La: $1s^2 2s^2 2p^6 3s^2 3p^6 4s^2 3d^{10} 4p^6 5s^2 4d^{10} 5p^6 6s^2 5d^1$

8. a) representative
   b) rare gas
   c) representative
   d) transition
   e) rare earth
   f) rare gas

9. a) N has 7 electrons. Their quantum numbers are:

	n	ℓ	m	s
electron 1	1	0	0	$+\frac{1}{2}$
2	1	0	0	$-\frac{1}{2}$
3	2	0	0	$+\frac{1}{2}$
4	2	0	0	$-\frac{1}{2}$
5	2	1	-1	$+\frac{1}{2}$
6	2	1	0	$+\frac{1}{2}$
7	2	1	+1	$+\frac{1}{2}$

b) Mg has 12 electrons. Their quantum numbers are:

	n		m	s
electron 1	1	0	0	$+\frac{1}{2}$
2	1	0	0	$-\frac{1}{2}$
3	2	0	0	$+\frac{1}{2}$
4	2	0	0	$-\frac{1}{2}$
5	2	1	-1	$+\frac{1}{2}$
6	2	1	-1	$-\frac{1}{2}$
7	2	1	0	$+\frac{1}{2}$
8	2	1	0	$-\frac{1}{2}$
9	2	1	+1	$+\frac{1}{2}$
10	2	1	+1	$-\frac{1}{2}$
11	3	0	0	$+\frac{1}{2}$
12	3	0	0	$-\frac{1}{2}$

10. a) S has 16 electrons.

$1s^2 \quad 2s^2 \quad 2p^6 \quad 3s^2 \quad 3p^4$

b) K has 19 electrons.

$1s^2 \quad 2s^2 \quad 2p^6 \quad 3s^2 \quad 3p^6 \quad 4s^1$

c) N has 7 electrons

$1s^2 \quad 2s^2 \quad 2p^6$

11. The subshell with the lowest value of $(n + \ell)$ is occupied by electrons first. If two subshells have the same $(n + \ell)$ value, the one with the lowest n value is

186 The Electronic Structure of Atoms

occupied by electrons first.

a) 5s: n = 5
    ℓ = 0
    ―――
    5

4d: n = 4
    ℓ = 2
    ―――
    6

So, since 5 is less than 6, the 5s subshell is occupied by electrons first.

b) 5s: n = 5
    ℓ = 0
    ―――
    5

4p: n = 4
    ℓ = 1
    ―――
    5

Since the (n + ℓ) values are the same, we know that the 4p subshell is occupied by electrons before the 5s because the 4p subshell has the lower n value.

c) 5d: n = 5
    ℓ = 2
    ―――
    7

4f: n = 4
    ℓ = 3
    ―――
    7

Both have the same (n + ℓ) value, so the 4f subshell is occupied by electrons before the 5d because the 4f subshell has the lower n value.

ANSWERS TO MULTIPLE CHOICE EXERCISES

1.	b	13.	c
2.	b	14.	b
3.	c	15.	a
4.	c	16.	d
5.	b	17.	c
6.	a	18.	b
7.	b	19.	b
8.	b	20.	c
9.	a	21.	c
10.	c	22.	b
11.	c	23.	b
12.	c	24.	a

CHAPTER 10
Chemical Bonding

REVIEW OF CHAPTER OBJECTIVES

1. Determine the number of valence electrons a
    representative element has (given its location in the
    periodic table or its electron configuration) and write
    an electron-dot structure for the element (section 10.2).

    Valence electrons are the electrons in the outermost
    electron shell of representative elements, the shell with
    the highest shell number (n). Note that this definition
    applies only to representative elements. Valence
    electrons will always be found in s and p subshells.

    Example: How many valence electrons are in an atom of:

    > a) K   b) P   c) I

Solution:

> a) K has 1 valence electron, in the 4th shell.
>
> K $1s^2 2s^2 2p^6 3s^2 3p^6 4s^1$
>
> b) P has 5 valence electrons, in the 3rd shell.
>
> P $1s^2 2s^2 2p^6 3s^2 3p^3$
>
> c) I has 7 valence electrons, in the 5th shell.
>
> I $1s^2 2s^2 2p^6 3s^2 3p^6 4s^2 3d^{10} 4p^6 5s^2 4d^{10} 5p^5$

An electron dot structure consists of the symbol for the
element and a dot for each valence electron. All
representative elements in the same group have the same
number of valence electrons, and the number of valence
electrons of these elements equals the group number (on
the periodic table).

IA	IIA	IIIA	IVA	VA	VIA	VIIA

Li• •Be•   •B•   •C•   •N: :O: :F:

Example: Draw electron-dot structures for:

   a) K   b) P   c) I

Solution:

From the electron configurations (as given in the previous example) we can write, directly, the electron-dot structures.

   a) K•   b) •P: c) :I:

2. <u>State the octet rule, and the basis for the rule</u> (section 10.3).

A key concept in the modern theory of chemical bonding is that certain arrangements of valence electrons are more stable than others.

The valence electron configurations of the rare gases, He, Ne, Ar, Kr, Xe and Rn are the most stable, as indicated by the fact that rare gases are the most unreactive of the elements. The octet rule states that <u>in compound formation atoms lose, gain or share electrons to attain a rare-gas electron configuration for each of the atoms involved</u>. All rare gases, except He, have eight electrons in their outer shell. Therefore, atoms will lose, gain or share electrons to obtain eight electrons in their outer shell, except when the outer shell is the first shell (which is filled with only two electrons, as in He).

3. <u>Identify by name and symbol monoatomic ions that are isoelectronic with a given rare gas, and write the electron configuration of these ions</u> (section 10.3).

Ionic compounds usually contain metallic and nonmetallic elements. The metallic element (in the compound) has lost electrons to form a positive ion. The nonmetallic element in the compound has gained electrons to form a negative ion. These ions are isoelectronic with rare gas atoms. Isoelectronic species contain the same number of electrons.

Example: Show how many electrons are lost or gained when the following atoms form ions that are isoelectronic with a rare gas:

a) F   b) Na   c) O   d) Mg

Solution:

a) The condensed electron configuration of F is:

$2e^-$ $7e^-$

The F atom gains 1 electron, to become a F^- ion,

$2e^-$ $8e^-$

which is isoelectronic with Ne ($2e^-$ $8e^-$)

b) The condensed electron configuration of Na is:

$2e^-$ $8e^-$ $1e^-$

The Na atom loses 1 electron to become a Na^+ ion, which is isoelectronic with Ne ($2e^-$ $8e^-$)

c) The condensed electron configuration of O is:

$2e^-$ $6e^-$

The O atom gains two electrons to become a O^{2-} ion, which is isoelectronic with Ne ($2e^-$ $8e^-$)

d) The condensed electron configuration of Mg is:

$2e^-$ $8e^-$ $2e^-$

The Mg atom loses two electrons to become a Mg^{2+} ion, which is isoelectronic with Ne ($2e^-$ $8e^-$)

Although isoelectronic species have identical electron configurations, they are not identical. In the example above, all the isoelectronic species have 10 electrons; however, each species has a different number of protons.

The number of electrons lost or gained by atoms during compound formation can be related to periodic table position of the elements involved.

Group IA metals, which are one periodic table position past a rare gas, lose 1 electron to form +1 ions.

Group IIA metals which are 2 periodic table positions past a rare gas, lose 2 electrons to form +2 ions.

Group IIIA metals which are 3 periodic table positions past a rare gas, lose 3 electrons to form +3 ions.

Group VIIA nonmetals which are 1 periodic table position before a rare gas, gain 1 electron to form −1 ions.

Group VIA nonmetals which are 2 periodic table positions before a rare gas, gain 2 electrons to form −2 ions.

Group VA nonmetals, which are 3 periodic table positions before a rare gas, gain 3 electrons to form −3 ions.

(Group IVA elements, equidistant between two rare gases, usually do not form ions.)

4. <u>Use electron-dot structures to describe the bonding in simple ionic compounds</u> (section 10.4).

Electron-dot structures can be used to illustrate ionic compound formation via electron transfer.

Example: Show the formation of the following ionic compounds using electron-dot structures:

a) KI   b) MgS   c) K_2S d) MgI_2

Solution:

a) Iodine needs one more electron which is supplied by potassium

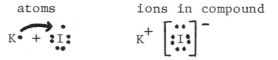

b) Sulfur needs two more electrons; magnesium has two excess electrons which are transferred

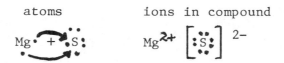

c) Sulfur needs two additional electrons. A potassium atom has only one electron to donate. Therefore, two potassium atoms are needed to meet the needs of one sulfur atom.

atoms   ions in compound

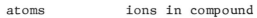

d) Magnesium has two electrons available for transfer.
   Iodine atoms can only accommodate one additional
   electron. Therefore, 2 iodine atoms are needed
   per magnesium atom.

atoms   ions in compound

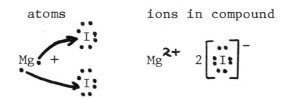

5. <u>Use electron-dot structures to describe the bonding in
    simple covalent compounds</u> (section 10.5 and 10.7).

In binary ionic compounds, the two atoms are dissimilar
(a metal and a nonmetal). Covalent compounds form
between similar or identical atoms; both are usually
nonmetals.

The simplest covalent bonding occurs between two like
atoms that complete their outer shells by sharing 1
electron from each atom.

<pre>
              atoms   molecule

    H₂   H. + ˙H ⟶ H:H
                ..   ..
    F₂   :F˙ + .F: ⟶ :F:F:
                ..   ..
</pre>

The pair of electrons shown between the two atoms in the
molecule are shared by both atoms. Each H atom in H_2
has a share of two electrons, and is isoelectronic with
the rare gas helium. Each F atom in F_2 has a share in
eight electrons, and is isoelectronic with the rare gas
neon.

Two different nonmetallic atoms can also share electrons
or form a covalent bond. The compound HF results when H
and F share electrons.

<pre>
              atoms   molecule
                          ..   ..
         H. + ˙F: ⟶ H:F:
                          ..   ..
</pre>

The number of covalent bonds that an atom forms is equal to the number of electrons it needs to become isoelectronic with a rare gas.

Example: Write electron dot structures for the simplest binary compound of:

a) P and Br   b) C and F   c) O and Cl

Solution:

a) P has 5 valence electrons and must form 3 covalent bonds. Br has 7 valence electrons and may only form 1 covalent bond.

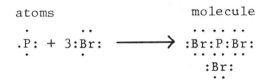

b) C has 4 valence electrons, and must form 4 covalent bonds. F has 7 valence electrons and may form only 1 covalent bond.

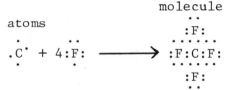

c) O has 6 valence electrons and must form 2 covalent bonds. Cl has 7 valence electrons and may form only 1 covalent bond.

atoms   molecule

$$\overset{\cdot}{:}\underset{\cdot}{O}: \; + \; 2:\overset{\cdot\cdot}{\underset{\cdot}{Cl}}: \; \longrightarrow \; :\overset{\cdot\cdot}{Cl}:\overset{\cdot\cdot}{O}:$$
$$:\underset{\cdot\cdot}{Cl}:$$

6. <u>Understand the relationship between the magnitude of an element's electronegativity and the position of an element on the periodic table</u> (section 10.6).

Although it is convenient to speak of a bond as ionic or covalent most chemical bonds are not ideal (they are not 100% ionic or 100% covalent). Between the two extremes of transfer of electrons and equal sharing of electrons are many cases of unequal sharing of electrons.

Unequal sharing of electrons results from some atoms
having a greater ability than others to attract shared
electrons to themselves.

A measure of the unequal sharing of electrons in a bond
is indicated by the electronegativities of the atoms
sharing the electrons. Electronegativity is the
attractive force that an atom in a molecule has for
shared electrons. Electronegativity values have no
units, and they range from 4.0 (for F, the most
electronegative element) to 0.7 for Fr (the least
electronegative element known). The higher the
electronegativity, the greater the electron-attracting
ability of the element.

For the representative elements, electronegativity
increases from left to right across a period of the
periodic table, and also increases from bottom to top in
a group (family) of the periodic table.

7. Classify, using a table of electronegativities, a given
   bond as nonpolar covalent, polar covalent or ionic
   (section 10.6).

   Bonds between identical atoms (which therefore have no
   difference in electronegativity) are nonpolar. For
   example, the bond in H_2 or Cl_2 is called a nonpolar
   covalent bond.

   Bonds between atoms whose difference in electronegativity
   is greater than zero but less than 1.7 are called polar
   covalent bonds.

   Bonds between atoms whose difference in electronegativity
   is greater than 1.7 are called ionic bonds.

   Example: Indicate whether the following bonds are polar
   covalent, nonpolar covalent or ionic:

   a) Br_2 b) HBr   c) CsBr

   Solution:

   a) Both Br atoms have an electronegativity of 2.8. A
      bond between identical atoms is nonpolar covalent;
      the electrons are shared equally.

   b) The difference in electronegativity between Br and
      H is 2.8 - 2.1 = 0.7. The bond is a polar
      covalent bond.

c) The difference in electronegativity between Br and Cs is 2.8 - 0.7 = 2.1. The bond is ionic.

8. <u>Understand what is meant by the terms single bond, double bond, triple bond and multiple bond</u> (section 10.7).

Covalent bonds can involve one or more shared pairs of electrons. A single shared pair of electrons is called a single covalent bond. The term multiple bond means a bond composed of more than one shared pair of electrons, and includes double bonds (where 2 pairs of electrons are shared between two atoms) and triple bonds (where 3 pairs of electrons are shared between two atoms). Triple bonds are stronger than double bonds, and double bonds are stronger than single bonds.

Only single bonds can form to an atom needing 1 electron to complete its outer shell. Atoms needing 2 or more electrons to complete the outer shell may form double bonds. Atoms needing 3 or more electrons to complete the outer shell may form triple bonds. A triple bond, 3 shared pairs of electrons, is the largest number of electrons 2 atoms of representative elements can share between them.

Example: Draw electron dot structures for the following and state the number of single, double and triple bonds in each molecule:

a) HCl b) H_2O c) CH_4 d) C_2H_2 e) C_2H_4

Solution:

Arrange atoms with all electrons in single bonds, then rearrange electrons to form double bonds, or triple bonds if single bonds don't complete the outer shells of all the atoms in the molecule.

a) H:Cl: 1 single bond

b) H:O: 2 single bonds
    H

c)   H
   H:C:H   4 single bonds
   H

d) H:C:::C:H   2 single bonds and 1 triple bond

```
        H  H
        ··  ··
  e)   C::C        4 single bonds and 1 double bond
        ··  ··
        H  H
```

Carbon can form single, double or triple bonds. However, each carbon atom must form 4 bonds (4 single bonds, 2 single bonds and 1 double bond, or 1 single bond and 1 triple bond).

9. Understand what is meant by a coordinate covalent bond (section 10.8).

A coordinate covalent bond is a bond in which both electrons of a shared pair come from 1 of the 2 atoms sharing the electrons (rather than having 1 electron from each atom).

In the compound chlorous acid, $HClO_2$, there is one coordinate covalent bond.

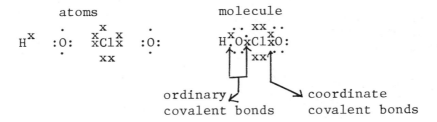

Although both electrons of a coordinate covalent bond come from one atom, once a coordinate covalent bond forms it is no different than any other covalent bond.

10. Determine when resonance structures are needed to describe the bonding in a molecule (section 10.9).

In some molecules, when no single electron-dot structure can be written which is consistent with bond strengths and bond distances, 2 or more resonance structures are used to represent the bonding. Resonance structures are 2 or more electron-dot structures for a molecule or an ion which have the same arrangement of atoms and contain the same number of electrons, and differ only in the location of the electrons.

The resonance structures for the nitrate ion, NO_3^{-1}, are:

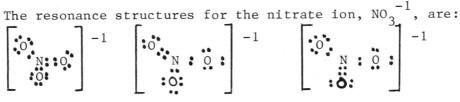

The actual bonding in the NO_3^{-1} ion is a resonance hybrid of these three structures, that is, an "average" of that depicted by the resonance structures.

11. <u>Write an electron-dot structure for a complicated molecule (with many electrons) or a polyatomic ion using a systematic approach</u> (section 10.10).

Trial and error determination of the electron dot structure of complicated molecules or polyatomic ions may not be the best approach. For any molecule or polyatomic ion which obeys the octet rule (not all do, but we'll consider only those that do) the following guidelines offer a useful procedure.

Step 1. Determine the arrangement of the atoms in the molecule (or ion). The central atom is usually the atom appearing only once in the formula, the C in CO_2 or N in NH_3. In oxyacids, the atom other than H or O is the central atom (S in H_2SO_3, and Cl in $HClO_3$). C is the central atom in almost all ternary C containing compounds; H and F are never central atoms.

Step 2. Determine the total number of valence electrons present by adding the number of valence electrons in each atom (and adding or subtracting the required number of electrons in negative or positive ions).

Step 3. Determine the total number of valence electrons needed by the molecule or ion for each atom to possess a rare gas electron configuration.

Step 4. Determine the number of bonding electrons needed (the number of shared electron pairs needed).

$$\text{\# bonding electrons pairs} = \frac{\substack{\text{\# electrons} \\ \text{needed} \\ \text{(step 3)}} - \substack{\text{\# electrons} \\ \text{present} \\ \text{(step 2)}}}{2}$$

Step 5. Arrange the bonding electron pairs around the atoms, then add the nonbonding electron pairs as needed to satisfy the octet rule.

Step 6. Check to see that the electron-dot structure contains the same number of electrons as that calculated in step 2, and also check that each atom satisfies the octet rule.

Example: Write electron-dot structures for:

  a) the oxyacid $HClO_4$ b) the molecule CO

Solution:

a) **Step 1:** The central atom is Cl, and the arrangement of atoms is:

```
        O

    O   Cl  O   H

        O
```

Step 2:
$$Cl \text{ has } 7 \text{ valence electrons: } \quad 1 \times 7 = 7$$
each O has 6 valence electrons: $6 \times 4 = 24$
H has 1 valence electron: $\quad 1 \times 1 = \underline{\quad 1}$

# of electrons = 32

Step 3: The number of electrons needed for rare gas electron configurations for each atom are:

$$1\ Cl = 1 \times 8 = \quad 8$$
$$4\ O\ = 4 \times 8 = 32$$
$$1\ H\ = 1 \times 2 = \underline{\quad 2}$$

42 electrons

Step 4: The number of electron pairs shared

$$= \frac{42-32}{2} = 5 \text{ shared electron pairs}$$
$$= 5 \text{ bonding electron pairs}$$

Step 5: There are 5 bonding locations, and 5 bonding electron pairs, so each bond is a single bond.

```
        O
        ••
    O:Cl:O:H
        ••
        O
```

Adding nonbonding electron pairs gives:

```
       ••
      :O:
       ••
    •• •• ••
   :O:Cl:O:H
    •• •• ••
      :O:
       ••
```

Step 6: All atoms have rare gas electron configurations, and the total number of electrons is 32 (as in step 2).

b) **Step 1:** The arrangement of atoms is:

C   O

Step 2: C has 4 valence electrons and oxygen has 6 valence electrons = 10 electrons.

Step 3: The number of electrons needed for each atom to have a rare gas electron configuration is:

C = 1 x 8 = 8
O = 1 x 8 = 8

16 electrons

Step 4: The number of bonding electron pairs

$$= \frac{16-10}{2} = \frac{6}{2} = 3$$

Step 5: With 3 bonding electron pairs, and only 1 bonding area, there must be a triple bond:

C:::O

Adding the nonbonding electron pairs gives:

:C:::O:

Step 6: Each atom has an octet of electrons, and there are 10 electrons (shown in step 2).

12. <u>Given its geometry, determine whether a molecule is polar or nonpolar</u> (section 10.11).

All covalent bonds between unlike atoms are polar; the greater the difference in electronegativity, the more polar the bond.

Molecules containing polar bonds may or may not be polar, depending on the molecular geometry (arrangement in space) of the atoms in the molecule.

With molecules containing 3 or more atoms (2 or more bonds) the collective effect of individual bond polarities determines molecular polarity. Sometimes the individual bond polarities cancel (due to the geometry of the molecules); other times they do not.

Example: Determine the polarity of each of the following molecules:

    a) PCl_3 (trigonal pyramid)

    b) CCl_3H (tetrahedral)

    c) H_2Se (angular)

    d) CBr_4 (tetrahedral)

Solution:

    a) The bond polarities in the trigonal pyramid PCl_3 do not cancel, so the molecule is polar.

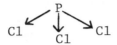

    b) In the tetrahedral CCl_3H molecule, the electron density is shifted toward the Cl and away from the H, so the molecule is polar.

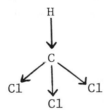

    c) In the angular H_2Se molecule, the shift in electron density is toward the Se and away from the H, because Se is more electronegative than H, and the molecule is polar.

    d) In the tetrahedral CBr_4 molecule, although each C to Br bond is polar (because Br is more electronegative than C), the molecule is symmetrical and therefore the molecule is nonpolar.

13. Optional. <u>Determine the geometry of e molecule using VSEPR theory</u> (section 10.12).

The basic principle behind valence shell electron pair

repulsion theory (VSEPR theory, for short) is that: the sets of valence shell electron pairs about an atom (bonding or nonbonding; single, double or triple bonds) tend to orient themselves so as to minimize repulsions between them (all electrons are negatively charged, and like charges repel each other).

Two sets of electron pairs are found as far as they can get from each other (at 180° angles to each other), which is a linear arrangement:

$$Z —— Y —— Z$$

Three sets of electron pairs, to get as far apart as possible, will be at 120° angles, a trigonal planar arrangement.

X   X
 \\   /
   Y
   |
   X

Four sets of electron pairs, to get as far apart as possible, will be at 109° angles, a tetrahedral arrangement.

A set of electrons may be: (1) a nonbonding electron pair, or (2) a single covalent bond (an electron pair), or (3) a double covalent bond (two electron pairs) or (4) a triple covalent bond (three electron pairs).

The steps in using VSEPR theory to determine molecular geometry are:

Step 1: Draw the elctron dot structure for the molecule or polyatomic ion. Since the electron dot structure is 2 dimensional, it doesn't directly show the geometry.

Step 2: Determine, from the electron dot structure, the number of sets of electron pairs around the central atom.

Step 3: Determine the arrangement of the electron sets

about the central atom that minimizes electron repulsions.

4 electron sets = tetrahedral arrangement
3 electron sets = trigonal planar arrangement
2 electron sets = linear arrangement

Step 4:   Describe the shape of the molecule (or polyatomic ion) in terms of the positions of the atoms. The molecular shape may be the same as, or different from, the electron set geometry. When there are no nonbonding electron pairs, the molecular shape is the same as the electron set geometry; when there are nonbonding electron pairs, the molecular shape is different from the electron set geometry.

In NH_3,

there are 4 electron sets which are arranged in a tetrahedral arrangement. However, the molecular geometry of the N and the 3 H atoms is trigonal pyramidal.

With 4 sets of electron pairs, the electron set geometry is tetrahedral. The molecular geometry is also tetrahedral if all 4 electron sets are bonding. It is trigonal pyramidal if 1 electron pair is nonbonding, and it is angular if there are 2 nonbonding electron pairs.

With 3 sets of electron pairs the electron set geometry is trigonal planar. If there are no nonbonding electron pairs, the molecular geometry is also trigonal planar, but if there is a nonbonding electron pair, the molecular geometry is angular.

With 2 sets of electron pairs, both electron set geometry and molecular geometry are linear.

Example: What is the molecular shape of:

    a) PH_3 b) CH_4

Solution:

    a) Step 1:   The electron dot structure for PH_3 is:

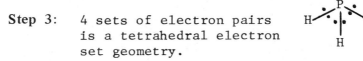

Step 2: The central atom, P, has 4 sets of electron pairs.

Step 3: 4 sets of electron pairs is a tetrahedral electron set geometry.

Step 4: There is 1 nonbonding electron pair, so the molecular geometry (configuration of the atoms) is trigonal pyramidal.

b) CH_4, as in part a) above, has 4 electron sets. However, since there are no nonbonding electron pairs, both the electron set geometry and the molecular geometry is tetrahedral.

PROBLEM SET

1. Draw electron-dot structures for atoms of:

   a) Sr   b) Al   c) Te

2. Predict the charge on the monoatomic ion that would be formed by:

   a) Sr   b) Te   c) Cs

3. Write electron-dot structures for the following ions:

   a) Cl^- b) Na^+ c) Br^- d) Mg^{2+} e) N^{3-}

4. Use electron-dot structures to describe the bonding in the following ionic compounds:

   a) K_2S b) CaI_2

5. Draw electron dot structures for the following covalent molecules, all of which contain only single bonds:

   a) Br_2 b) NH_3 c) HI   d) C_2H_6

6. Draw electron-dot structures for the following covalent molecules, all of which contain multiple bonds:

   a) CO   b) CO_2 (O C O)   c) C_2Cl_2 (Cl C C Cl)

7. Draw electron-dot structures for the following polyatomic ions:

   a) IO_3^- b) ClO_2^- c) BrO_4^- d) NH_4^+

8. Draw electron-dot structures for the following molecules:

   a) $POCl_3$ (the central atom is P)

   b) $COCl_2$ (the central atom is C)

   c) CH_2O (the central atom is C)

9. Define electronegativity.

Optional Problems

10. Using VSEPR theory, determine the molecular geometry of the following molecules:

    a) OF_2 b) SO_3 c) CF_4

11. Using VSEPR theory, determine the molecular geometry of the following polyatomic ions:

    a) ClO_3^- c) OCN^- (cyanate ion; C is the central
    b) SO_4^{2-} atom)
                              d) ClO_2^-

MULTIPLE CHOICE EXERCISES

1. How many valence electrons are there in the

representative element with the electron configuration $1s^2 2s^2 2p^4$?

a) 2
b) 4
c) 6
d) 8

2. For which of the following elements is the listed number of valence electrons incorrect?

a) Al -- 3 valence electrons
b) Cl -- 5 valence electrons
c) O -- 6 valence electrons
d) Si -- 4 valence electrons

3. Which of the following is a correct electron dot diagram for sulfur?

a) :S̈:

b) .Ṡ.

c) S̈:

d) .S̈:

4. The "octet rule" states:

a) that compounds comprised of 8 atoms are very stable
b) charges of 8+ and 8- are very stable
c) that atoms, during compound formation, strive to obtain 8 valence electrons
d) that there are 8 steps in writing the formula of an ionic compound

5. Elements in groups IIIA and VIIA of the periodic table would, respectively, be expected to form ions with charges of:

a) +3 and +7
b) -3 and -7
c) +3 and -1
d) -3 and +1

6. In which of the following sequences are all of the particles isoelectronic?

   a) F, Ne, Na

   b) Mg, Al, Si

   c) Cl^-, Ar, K^+

   d) O^{2-}, S^{2-}, Se^{2-}

7. In the formation of the ionic compound CaO the number of electrons transferred from Ca to O, per formula unit, is:

   a) 1
   b) 2
   c) 3
   d) 4

8. The correct electron-dot structure for the covalent molecule H_2O is:

   a) :H:Ö:
      ..
       H

   b) H:O
       ..
       H

   c) H:Ö:
      ..
       H

   d) :Ḧ:Ö:
      .. ..
       :H:
       ..

9. The total number of "dots" in the electron-dot structure for CO_2 is:

   a) 4
   b) 8
   c) 16
   d) 20

10. The most electronegative element on the electronegativity scale is:

    a) fluorine
    b) cesium
    c) hydrogen
    d) element 106

11. Which of the following sequences of elements is arranged in order of increasing electronegativity?

    a) Pb, Se, Ge, O
    b) Se, Pb, Ge, O
    c) Pb, Ge, Se, O
    d) O, Se, Pb, Ge

12. Given the hypothetical elements A, B, C and D with electronegativities of 3.2, 2.8, 1.2 and 2.5 respectively, determine which of the following bonds would be ionic:

    a) BD
    b) AD
    c) AC
    d) CD

13. Double bonds:

    a) may be ionic or covalent
    b) are found only in polyatomic ions
    c) occur only between atoms which have 2 valence electrons
    d) involve the sharing of four electrons between atoms

14. Coordinate covalent bonds differ from normal covalent bonds in that:

    a) three atoms are always involved rather than two
    b) they must always be triple bonds
    c) one atom supplies more electrons to the bond than the other atom
    d) the electrons of the bond move about in a coordinated pattern

15. One of the resonance structures for the molecule SO_2 is

    O〓S—O . How many other resonance structures are there?

    a) 1
    b) 2
    c) 3
    d) 4

16. Resonance structures describing bonding in a covalent molecule differ only in:

    a) arrangement of atoms
    b) location of the valence electrons with the molecule
    c) number of valence electrons present
    d) total number of bonds shown

17. The electron-dot structure for the PO_4^{3-} ion contains how many bonding electrons?

    a) 2
    b) 4
    c) 8
    d) 12

18. Which of the following is an example of a molecule which contains polar covalent bonds and yet is a nonpolar molecule?

    a) O–H
       |
       H

    b) O=C=O

    c) H–Cl

    d) N≡N

19. AB and AB_2 are both covalent compounds, where A and B are elements of differing electronegativity. AB_2 has an angular geometry. Which of the following statements concerning the polarity of these molecules is true?

    a) AB and AB_2 both must be polar molecules

    b) AB_2 must be nonpolar

    c) AB must be polar

    d) AB and AB_2 each can be polar or nonpolar

Optional

20. A premise of VSEPR theory is that 3 sets of electron pairs about a central atom (whether bonding or nonbonding) will be located in a:

    a) tetrahedral arrangement
    b) trigonal planar arrangement
    c) trigonal pyramidal arrangement
    d) linear arrangement

21. The shape (geometry) of a molecule, where the central atom has 2 bonding sets and one nonbonding set of electron pairs will be:

    a) linear
    b) angular
    c) trigonal planar
    d) trigonal pyramidal

22. What is the molecular shape of the NO_2^+ ion?

    a) linear
    b) angular
    c) trigonal pyramidal
    d) trigonal planar

ANSWERS TO PROBLEMS SET

1. a) Sr·   b) ·Al·   c) :Te·

2. a) Sr^{2+} b) Te^{2-} c) Cs^+

3. a) $[:Cl:]^-$ b) Na^+ c) $[:Br:]^-$ d) Mg^{2+}

    e) $[:N:]^{3-}$

4. a) K·   :S: $2K^+ + [:S:]^{2-} \longrightarrow K_2S$
      +   K·

    b)   :I: $Ca^{2+} + 2[:I:]^- \longrightarrow CaI_2$
     Ca· +
            :I:

5. a) :Br:Br: b) H:N:H   c) H:I: d) H H
                    H   H:C:C:H
                                              H H

6. a) :C::O: b) O::C::O   c) :Cl:C:::C:Cl:

7. a) $[:O:I:O:]^-$ b) $[:Cl:O:]^-$ c) $\begin{bmatrix} :O: \\ :O:Br:O: \\ :O: \end{bmatrix}^-$
          :O: :O:

    d) $\begin{bmatrix} H \\ H:N:H \\ H \end{bmatrix}^+$

8. a)   :O: b) :O: c) H
   :Cl: P :Cl: C:Cl: C::O:
      :Cl: :Cl: H

9. Electronegativity is the attractive force that an atom of an element has for shared electrons in a molecule.

10. a) :Ö:F̈:   angular   (2 bonding pairs, and 2
     :F̈:   nonbonding pairs results in an
                                          angular molecular geometry)

    b)   :O: trigonal planar   (3 electron sets and no
           ::   nonbonding electron pairs
           S   results in a trigonal planar
       :Ö. .Ö: molecular geometry)

    c)   :F̈:   tetrahedral   (4 electron sets and no
       :F̈:C:F̈:   nonbonding electron pairs
           :F̈:   results in a tetrahedral
                                             molecular geometry)

11. a) [:Ö:C̈l:Ö:]⁻   trigonal pyramidal, because it has 4
           : O :   electron sets, one of which is a
                               nonbonding electron pair.

    b) [:Ö:]²⁻   tetrahedral, having 4 electron sets,
         :Ö:S:Ö: none of which are nonbonding electron
           :Ö: pairs.

    c) [Ö::C::N̈]⁻   linear, having 2 electron sets (and no
                               nonbonding electron pairs)

    d) [:Ö:C̈l:]⁻   angular, having 4 electron sets, 2 of
           : O :   which are nonbonding electron pairs.

ANSWERS TO MULTIPLE CHOICE EXERCISES

1. c   12. c
2. b   13. d
3. d   14. c
4. c   15. a
5. c   16. b
6. c   17. c
7. b   18. b
8. c   19. c
9. c   20. b
10. a   21. b
11. c   22. a

CHAPTER 11
States of Matter

REVIEW OF CHAPTER OBJECTIVES

1. <u>Compare and contrast the gross distinguishing properties of gases, liquids and solids</u> (section 11.2).

   The obvious differences between solids, liquids and gases are that:

   > A solid has a definite volume and a definite shape.

   > A liquid has a definite volume, but takes the shape of the bottom of its container.

   > A gas has neither a definite volume nor a definite shape; it completely fills its container.

   Solids and liquids have:

   > High density, low compressibility and low thermal expansion.

   Gases have:

   > Low density, large compressibility and moderate thermal expansion.

2. <u>Understand the roles that kinetic energy (disruptive forces) and potential energy (cohesive forces) play in determining the physical state of a system</u> (section 11.3).

   The kinetic molecular theory is one of the fundamental theories of chemistry. Its basic idea is that matter is composed of particles in motion.

   The kinetic molecular theory applies to solids and liquids and gases. According to this theory:

   1. Matter is composed of tiny particles (atoms or molecules).

2. Particles possess kinetic energy because they are in constant random motion.

3. Particles possess potential energy because of attractive and repulsive forces between them.

4. The velocity of the particles increases with temperature.

5. Energy is transferred from particle to particle without any loss of energy.

Kinetic energy is energy a particle possesses due to its motion.

$$KE = 1/2 \ mv^2$$

Kinetic energy may be considered to be a disruptive force, tending to make particles increasingly independent of each other.

Potential energy is the energy a particle possesses because of forces of attraction or repulsion (between it and other particles). Potential energy may be considered to be a cohesive force tending to cause order among particles.

3. <u>Use kinetic molecular theory to explain the characteristic properties of solids, liquids and gases</u> (sections 11.3 through 11.6).

In the solid state (section 11.4) cohesive forces (potential energy) dominate over disruptive forces (kinetic energy). The cohesive forces in a solid draw the particles close together in a regular pattern (explaining why solids have high density, are only slightly compressible, have small thermal expansion and have a definite shape and a definite volume).

In the liquid state (section 11.5), neither potential energy (cohesive forces) not kinetic energy (disruptive forces) dominate. Particles are close together, but freely moving, and lack sufficient energy to separate from each other. The liquid has a definite volume and indefinite shape because attractive forces are strong enough to keep the particles together in a definite volume, but not strong enough to prevent the particles from sliding over one another. There is a high density and small compressibility and small thermal expansion (as in a solid) because the particles are close together.

In the gaseous state (section 11.6), kinetic energy (disruptive forces) dominates, and gas particles are essentially independent of each other. Therefore, the gas has an indefinite volume and an indefinite shape because there is little force of attraction between particles that are in rapid random motion. The low density and large compressibility of a gas results from the particles being far apart. Since velocity increases with temperature, gases show moderate thermal expansion.

4. <u>Contrast on a molecular level the differences between solids and liquids and gases</u> (section 11.7).

The gaseous state shows marked differences from the solid and liquid states, largely because a gas is mostly empty space. Since there are large spaces between gas molecules the gas has a low density, is easily compressed and undergoes significant thermal expansion. On the other hand, solids and liquids (which have particles closely packed together) have the similar characteristics of high density and little compressibility or thermal expansion.

5. <u>Understand the difference between exothermic and endothermic changes of state</u> (section 11.8).

An endothermic change of state requires the absorption of heat; heat must be used to melt, sublime or evaporate a substance.

An exothermic change of state is a change that results in the release of heat; heat is given off when a substance freezes or condenses.

6. <u>Explain, on a molecular basis, what happens during the process of evaporation, and list factors which affect the rate of evaporation of a liquid</u> (section 11.9).

Evaporation involves the escape of molecules from the surface of a liquid to form a vapor. The molecules of a liquid have varied energies, and those at the surface of the liquid with high enough energy escape by overcoming the attractive forces that hold molecules together in a liquid. Since evaporation results in the loss of high energy molecules from the surface of the liquid, the rate of evaporation of a liquid increases as temperature increases (because at higher temperature more molecules have sufficiently high energy to escape from the surface). Also, since high energy molecules leave the liquid, the temperature of the liquid decreases as evaporation proceeds.

7. <u>Understand the relationship between vapor pressure and an equilibrium state, and list the factors which affect the magnitude of the vapor pressure of a given liquid</u> (section 11.10).

A state of equilibrium is a situation in which two opposite processes occur at equal rates.

When a liquid evaporates in a closed container, an equilibrium state eventually occurs, in which liquid molecules leave the surface at the same rate as vapor molecules return to the liquid. At this point both evaporation and condensation occur at equal rates; a dynamic equilibrium of the two processes results. The vapor above the liquid exerts a constant pressure on the liquid surface and the container walls, called the vapor pressure. The vapor pressure of a liquid is the pressure exerted by the vapor above the liquid when the liquid and the vapor are in equilibrium. The magnitude of the vapor pressure of a liquid depends on the nature and temperature of the liquid. Nonvolatile liquids have strong attractive forces between molecules and have low vapor pressures (mercury is an example of a liquid with a low vapor pressure). Volatile liquids have weak attractive forces between molecules and have high vapor pressures (diethyl ether is an example of a liquid with a high vapor pressure).

8. <u>Understand the process of boiling from a molecular viewpoint; understand what is meant by the term normal boiling point, and know the relationship between the boiling point of a liquid and the external pressure</u> (section 11.11).

Boiling is a special form of evaporation in which the liquid changes to the vapor within the body of the liquid through the formation of bubbles.

The boiling point of a liquid is defined as the temperature at which the vapor pressure of the liquid is equal to the external pressure (atmospheric pressure) exerted upon it. The normal boiling point of a liquid is the temperature at which the liquid boils at a pressure of 760 torr (1 atm).

9. <u>Understand the difference between intramolecular forces</u>
    <u>and intermolecular forces, and describe and distinguish</u>
    <u>between the three types of intermolecular forces that</u>
    <u>may be operative in a liquid</u> (section 11.12).

    Intermolecular forces are forces which act <u>between</u> one
    molecule and another. The intermolecular forces of
    attraction between molecules in a liquid must be
    overcome in order for molecules to escape from the
    liquid into the vapor state.

    Intramolecular forces are the forces involved in covalent
    bonding and ionic bonding; intramolecular forces act
    <u>within</u> the molecule. Intramolecular forces are much
    stronger than intermolecular forces. However, despite
    the relative weakness of intermolecular forces, they
    greatly influence the behavior of liquids.

    The three principal types of intermolecular forces are:
    a) dipole-dipole interactions, b) hydrogen bonding and
    c) London forces.

    Dipole-dipole interactions exist between polar molecules
    (dipoles) which are electrically unsymmetrical. The
    greater the polarity of the molecules the stronger the
    dipole-dipole interactions.

    A special type of dipole-dipole interaction occurs
    between polar molecules which contain hydrogen bonded to
    a very electronegative element (F, O, N), which is called
    a <u>hydrogen bond</u>. The strongest hydrogen bond is the
    strongest of all intermolecular forces, and is about one-
    tenth as strong as a covalent bond. Hydrogen bonding
    plays an important part in the behavior of many
    biologically important compounds; complex biologically
    active compounds such as DNA, and "simple" compounds,
    such as H_2O.

    The third, and weakest, type of intermolecular force is
    the London force, which acts on all atoms and molecules.
    The strength of London forces depends on the ease with
    which electron distribution in an atom or a molecule can
    be distorted. If molecules are the same weight, the
    larger ones will have stronger London forces (because the
    outer electrons are farther from the nucleus).
    Heavier molecules have stronger London forces than
    (similar) lighter ones.

    The larger the London forces, the higher the boiling
    point of the liquid, since more energy is needed to break
    forces of attraction and allow liquid molecules to become

independent of each other. As examples, liquid CH_4 boils at $-162^{\circ}C$, while liquid CCl_4 boils at $77^{\circ}C$.

The London force is the most common type of intermolecular force; although it is the weakest of the intermolecular forces, it is the only attractive force between nonpolar molecules.

10. <u>Distinguish between crystalline and amorphous solids, and know the nature of the crystal lattice structure for each of the five types of crystalline solids</u> (section 11.13).

Solids fall into two cetegories: crystalline solids and amorphous solids. Most solids are crystalline solids, which are characterized by having a regular 3-dimensional arrangement of atoms, or ions, or molecules present. Examples include sodium chloride (ordinary table salt), and minerals such as quartz. Amorphous solids are characterized by having a random arrangement of the atoms or molecules present, such as paraffin.

Crystalline solids are much more common than amorphous solids; the word "solid" usually means crystalline solid.

The orderly arrangement of particles in a crystalline solid is called a crystal lattice, with the positions occupied by the particles in the crystal lattice being called crystal lattice sites. Crystalline solids are classified into five classes: a) ionic solids, b) polar molecular solids, c) nonpolar molecular solids, d) macromolecular solids and e) metallic solids.

Ionic solids have positive and negative ions at the crystal lattice sites. Each ion is surrounded by oppositely-charged nearest neighbors. Since interionic attractions are strong, ionic solids have high melting points and negligible vapor pressures.

Polar molecular solids have polar molecules at the lattice sites. The dipole-dipole and London forces which hold the molecules in the lattice positions are relatively weak, therefore polar molecular solids have moderate melting points and moderate vapor pressures.

Nonpolar molecular solids have molecules (or atoms for rare gases) at the lattice sites, held together only by London forces. Since London forces are the weakest of the intermolecular forces, nonpolar molecular solids have the lowest melting points and are volatile.

Macromolecular solids have atoms at the crystal lattice

sites. These atoms are bonded to their nearest neighbors by covalent bonds. Macromolecular solids, such as diamond, are not common, but have very high melting points and low volatility.

Metallic solids have metal atoms at crystal lattice sites. These atoms are thought to be held together in a "sea" of outer shell electrons. The movement of these free electrons accounts for the electrical conductivity of metals. Melting points of metallic solids vary; volatility is generally low.

11. <u>Calculate the heat released or absorbed when a substance is heated, cooled or undergoes a change in state (given the specific heat, heat of vaporization, and heat of fusion of the substance)</u> (section 11.14).

For a given substance, gas molecules contain more energy than solid molecules. We know this because a solid turns to a liquid when heated, and a liquid turns to a gas when heated (provided no decomposition occurs instead).

A heating curve shows the relationship of energy to the states of matter:

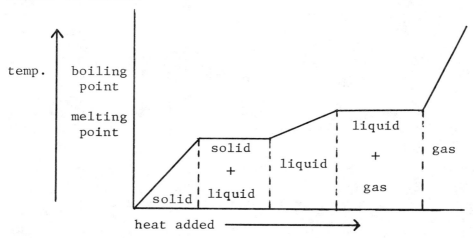

Note how a solid increases in temperature as heat is added, until it begins to melt; the temperature then remains unchanged as the solid becomes a liquid; the liquid then increases in temperature as it is heated, until it begins to vaporize (boil); the temperature of the liquid then remains unchanged as the liquid changes to a gas; and finally the temperature of the gas increases as more heat is added.

The actual amount of heat needed for each change varies from one substance to another.

Specific heat is defined as the heat required to change a given amount of a substance in a particular state by $1^\circ C$. The higher the specific heat of a substance, the smaller the change in temperature when a given amount of heat is applied.

The specific heat of liquid water is 1.00 cal/g$^\circ$C.

The specific heat of liquid Hg is 0.033 cal/g$^\circ$C.

Example: How much heat is required to change 25 grams of liquid water from $30^\circ C$ to $85^\circ C$?

Solution:

Heat absorbed = sp. ht. x mass x temperature change

$$= \left(\frac{1.00 \text{ cal}}{g^\circ C}\right) \times \left(25 \text{ g}\right) \times \left[(85 - 30)^\circ C\right]$$

= 1.00 cal x 25 x 55 = 1375 cal
(calculator answer)

= 1400 cal
(correct answer)

Heat energy is absorbed or released in changes of state. The heat of fusion is the amount of heat energy required to change one gram of a solid to a liquid at its melting point. The heat of vaporization is the amount of heat energy required to change one gram of a liquid to a gas at its boiling point. For water, the heat of fusion is 79.8 cal/g, and the heat of vaporization is $54\overline{0}$ cal/g.

Example: Calculate the heat required to change $8\overline{0}0$ grams of water at $100^\circ C$ from liquid to vapor.

Solution: ht. required = ht. vap. x mass

$$= \left(\frac{54\overline{0} \text{ cal}}{g}\right) \times \left(8\overline{0}0 \text{ g}\right)$$

= 432000 cal (calculator and correct answer)

Example: Calculate the heat required to change $11\overline{0}$ g of solid water (ice) at $0^\circ C$ to a liquid at $0^\circ C$.

Solution: ht. required = ht. fusion x mass

$$= \left(\frac{79.8 \ cal}{g}\right) \times \left(11\overline{0} \ g\right)$$

= 8778 cal (calculator answer)
= 8780 cal (correct answer)

Example: Calculate the heat released when 25.0 g of steam at 123°C changes to ice at −12.0°C.

Solution:

This problem can be broken down into 6 steps:

Step 1.   Steam at 123°C to steam at 100°C
(sp. ht. steam = 0.48 cal/g)

Step 2.   Steam at 100°C to water at 100°C
(ht. vap. = 54$\overline{0}$ cal/g)

Step 3.   Water at 100°C to water at 0°C
(sp. ht. water = 1.00 cal/g)

Step 4.   Water at 0°C to ice at 0°C
(ht. fus. = 79.8 cal/g)

Step 5.   Ice at 0°C to ice at −12.0°C
(sp. ht. ice = 0.50 cal/g°C)

Step 6.   Add steps 1–5

Step 1:   Steam from 123°C to 100°C

ht released = sp ht x mass x temperature change

$$= \left(\frac{0.48 \ cal}{g \ ^{\circ}C}\right) \times \left(25.0 \ g\right) \times \left[(123-100)\,^{\circ}C\right]$$

= 276 cal (calculator answer)
= 280 cal (correct answer)

Step 2:   Steam at 100°C to water at 100°C

ht released = ht vap x mass $= \left(\frac{54\overline{0} \ cal}{g}\right) \times \left(25.0 \ g\right)$

= 13500 cal (calculator and
            correct answer)

Step 3:   Water at 100°C to water at 0°C

ht released = sp ht x mass x temperature change

States of Matter   219

$$= \left(\frac{1.00 \text{ cal}}{\text{g} {}^{\circ}\text{C}}\right) \times \left(25.0 \text{ g}\right) \times \left[(100-0) {}^{\circ}\text{C}\right]$$

= 2500 cal (calculator answer)
= 2500 cal (correct answer)

Step 4: Water at $0{}^{\circ}$C to ice at $0{}^{\circ}$C

ht released = ht fusion x mass $= \left(\frac{79.8 \text{ cal}}{\text{g}}\right) \times \left(25.0 \text{ g}\right)$

= 1995 cal (calculator answer)
= 2000 cal (correct answer)

Step 5: Ice at $0{}^{\circ}$C to ice at $-12.0{}^{\circ}$C

ht released = sp ht x mass x temperature change

$$= \left(\frac{0.50 \text{ cal}}{\text{g} {}^{\circ}\text{C}}\right) \times \left(25.0 \text{ g}\right) \times \left[0-(-12) {}^{\circ}\text{C}\right]$$

= 150 cal (calculator and
correct answer)

Step 6: Add results of steps 1-5

(280 + 13500 + 2500 + 2000 + 150) cal = 18430 cal
(calculator answer)
= 18400 cal
(correct answer)

PROBLEM SET

1. Compare solids and liquids and gases in terms of:
a) volume and shape, b) density and compressibility and thermal expansion.

2. Explain the difference between crystalline solids and amorphous solids.

3. Given the following data tell which substances are solids, which are liquids and which are gases at:

a) room temperature ($22{}^{\circ}$C)
b) $100{}^{\circ}$C
c) $-50{}^{\circ}$C

substance	melting point	boiling point
Hg	-39.9 °C	357 °C
Cs	28.5 °C	670 °C
Ga	29.8 °C	1983 °C
Na	97.5 °C	880 °C
P	44.1 °C	280 °C
Br_2	-7.2 °C	58.8°C
CH_4	-184 °C	-161 °C
NH_3	-77.7 °C	-33.4°C
C_6H_6	5.51°C	80.1°C

4. Distinguish between the terms:

   a) kinetic energy and potential energy
   b) boiling and evaporation
   c) endothermic changes and exothermic changes

5. Using the data given in figure 11.1 (page 275) and Table 11.7 (page 294) of the textbook, and additional data given below, calculate:

   a) the heat needed to change 50.0 g of water at 25.0°C to steam at 125°C.

   b) the heat needed to warm $\overline{30}$ gallons of water (in a hot water tank) from room temperature (21°C, which is 70°F) to 54°C (which is 120°F)

   c) the heat released when 100.0 grams of molten copper, at its melting point of 1083°C, changes to solid copper at a temperature of 25°C. (Ht. fusion of Cu = 42 cal/g)

6. Given:

element	melting point	heat of fusion	specific heat of solid
Al	658°C	76.8 cal/g	0.28 cal/g°C
Pb	327°C	5.86 cal/g	0.036 cal/g°C
Au	1064°C	15.8 cal/g	0.03 cal/g°C

Calculate the amount of heat needed to change:

a) $\overline{500}$ g of Al at 600°C to liquid at its melting point

b) 1.00 kg of lead at 300°C to liquid at its melting point

c) a 10.0 g nugget of gold at room temperature (21°C) to liquid at its melting point

MULTIPLE CHOICE EXERCISES

1. Which of the following is not a property of both liquids and solids?

   a) high density
   b) small degree of compressibility
   c) small degree of thermal expansion
   d) definite shape

2. According to kinetic molecular theory, in most solids:

   a) there is no motion of the particles
   b) there is usually some vibrational movement of the particles
   c) the particles generally revolve around each other in a regular manner
   d) the particles are packed together in a rather haphazard manner

3. In the solid state cohesive forces are:

   a) roughly of the same magnitude as disruptive forces
   b) very weak compared to disruptive forces
   c) dominant over disruptive forces
   d) nonexistent

4. A liquid maintains its volume, whatever the shape or size of the container because:

   a) most of the volume of a liquid is empty space
   b) cohesive forces are strong enough to prevent molecules from separating from each other
   c) the particles are widely separated from each other
   d) cohesive forces are strong enough to hold the particles in fixed positions

5. Which of the following statements is not consistent with kinetic molecular theory for gases?

   a) the molecules of a gas are relatively far apart
   b) between collisions, molecules travel in straight lines
   c) when pressure is applied the molecules themselves decrease in size
   d) the molecules of a gas move essentially independently of each other

6. Which of the following pairs of states are most similar in terms of distance between molecules?

   a) solid and liquid
   b) solid and gas
   c) liquid and gas
   d) Utah and Hawaii

7. In which of the following pairs of changes are both changes exothermic?

   a) sublimation, evaporation
   b) freezing, melting
   c) freezing, condensation
   d) melting, sublimation

8. Increasing the temperature of a liquid increases the rate of evaporation because at a higher temperature:

   a) molecules expand in size (becoming less dense)
   b) attractive forces are weaker
   c) more molecules have the minimum energy needed for escape
   d) kinetic energy of the molecules decreases

9. Indicate the missing word or phrase in the following definition of vapor pressure: "Vapor pressure is the pressure exerted by a vapor _____ its liquid phase."

   a) present above
   b) in the same container as
   c) in equilibrium with
   d) when separated from

10. When a liquid is placed in a closed container:

    a) evaporation ceases
    b) evaporation continues for a time, then stops
    c) a condition is attained in which condensation of the vapor is occurring at the same rate as evaporation of the liquid

d) the vapor pressure above the liquid becomes equal to one atmosphere

11. The normal boiling point of any liquid is defined as:

    a) the temperature at which the liquid boils
    b) $100^{\circ}C$
    c) the temperature at which the vapor pressure equals one atmosphere of pressure
    d) the temperature at which there is equilibrium

12. Intramolecular forces differ from intermolecular forces in that:

    a) they occur only in gases
    b) they are much stronger than intermolecular forces
    c) they occur between molecules rather than within molecules
    d) they are not electrostatic in origin

13. In a liquid, a dipole-dipole interaction would most likely result from the interaction of:

    a) 2 nonpolar molecules
    b) 2 polar molecules
    c) a polar and a nonpolar molecule
    d) an inter-molecule and an intra-molecule

14. Hydrogen bonds are:

    a) extremely strong intermolecular forces
    b) extremely weak intermolecular forces
    c) extremely strong intramolecular forces
    d) extremely weak intramolecular forces

15. In a covalent network solid the particles occupying the lattice sites are:

    a) atoms
    b) nonpolar molecules
    c) polar molecules
    d) positive and negative ions

16. In a nonpolar molecular solid the forces between particles are:

    a) attractions between oppositely charged ions
    b) London forces
    c) hydrogen bonds
    d) dipole-dipole attractions

17. Which of the following quantities is needed in calculating the amount of heat energy released as water is cooled from $80^\circ C$ to $60^\circ C$?

   a) specific heat of water
   b) heat of fusion
   c) heat of vaporization
   d) both specific heat and heat of fusion

18. Which of the following would be correct units for a heat of vaporization value?

   a) $cal/^\circ C$
   b) cal/g
   c) $g/^\circ C$
   d) $cal/g^\circ C$

ANSWERS TO PROBLEMS SET

1. Solids have: a) definite shape and definite volume; b) high density and small compressibility and very small thermal expansion.

   Liquids have: a) definite volume and indefinite shape; b) high density, low compressibility and small thermal expansion.

   The major difference between solids and liquids is that solids have a definite shape and liquids don't. Otherwise, they are very similar, having high density and small compressibility.

   Gases are distinctly different from both liquids and solids in that they have: a) neither definite shape nor volume; b) low density, large compressibility and moderate thermal expansion.

2. Crystalline solids are most common and have a regular arrangement of particles. Amorphous solids, which are not as common, have a random arrangement of particles.

3. a) At room temperature $(22^\circ C)$, the solids are: Cs, Ga, Na, P; the liquids are: Hg, Br_2, C_6H_6; the gases are: CH_4, NH_3

   b) At $100^\circ C$, none of the listed substances are solids. the liquids are: Hg, Cs, Ga, Na and P; the gases are: Br_2, CH_4, NH_3 and C_6H_6.

c) At -50°C, the solids are: Hg, Cs, Ga, Na, P, Br_2, C_6H_6; the only liquid is NH_3; the only gas is CH_4.

4. a) Kinetic energy is a disruptive force, energy due to motion, while potential energy is a cohesive force, energy of attraction and repulsion.

   b) Boiling takes place throughout the liquid when the temperature is high enough for the vapor pressure of the liquid to equal outside pressure. Evaporation occurs at the surface of a liquid at all temperatures.

   c) Endothermic changes require (absorb) heat, while exothermic changes release (give off) heat.

5. a) Calculating the heat needed to change water at 25°C to steam at 125°C requires 4 steps.

   Step 1: Liquid water at 25°C to liquid water at 100°C

   ht required = sp ht x mass x temperature change

   $$= \left(\frac{1.00 \text{ cal}}{g^\circ c}\right) \times \left(50.0 \text{ g}\right) \times \left[(100 - 25)^\circ c\right]$$

   = 3750 cal (calculator answer)

   = 3800 cal (correct answer)

   Step 2: Liquid water at 100°C to water vapor at 100°C

   ht required = ht vap x mass $= \left(\frac{540 \text{ cal}}{g}\right) \times \left(50.0 \text{ g}\right)$

   = 27000 cal (calculator answer)

   = 27$\overline{0}$00 cal (correct answer)

   Step 3: Water vapor at 100°C to water vapor at 125°C

   ht required = sp ht x mass x temperature change

   $$= \left(\frac{.48 \text{ cal}}{g^\circ c}\right) \times \left(50.0 \text{ g}\right) \times \left[(125 - 100)^\circ c\right]$$

   = 600 cal (calculator answer)

   = 6$\overline{0}$0 cal (correct answer)

Step 4: Add steps 1-3

$$(3800 + 27\overline{0}00 + 6\overline{0}0) \text{ calories} = 31400 \text{ cal}$$
$$\text{(calculator and}$$
$$\text{correct answer)}$$

b) Calculating the heat needed to warm $3\overline{0}$ gallons of water (density = 1.0 g/mL) from 21°C to 54°C requires 2 steps.

Step 1. $3\overline{0}$ gallons water = ? grams water

$$3\overline{0} \text{ gal} \times \left(\frac{4 \text{ qt}}{1 \text{ gal}}\right) \times \left(\frac{0.946 \text{ liter}}{1 \text{ qt}}\right) \times \left(\frac{1 \text{ mL}}{10^{-3} \text{ liter}}\right)$$

$$\times \left(\frac{1.0 \text{ g } H_2O}{1 \text{ mL } H_2O}\right)$$

$$= 113520 \text{ g } H_2O \text{ (calculator answer)}$$
$$= 110000 \text{ g } H_2O \text{ (correct answer)}$$

Step 2. Water from 21°C to 54°C

ht required = sp ht x mass x temperature change

$$= \left(\frac{1.00 \text{ cal}}{\text{g} \cdot ^\circ C}\right) \times \left(110,000 \text{ g}\right) \times \left[(54 - 21)^\circ C\right]$$

$$= 3630000 \text{ cal (calculator answer)}$$
$$= 3600000 \text{ cal (correct answer)}$$

c) Calculating the heat released when 100.0 g of molten Cu at its melting point (of 1083°C) changes to solid copper at 25°C requires 3 steps:

Step 1. Liquid copper at 1083°C to solid copper at 1083°C

ht released = ht fusion x mass $= \left(42 \text{ cal/g}\right) \times \left(100.0 \text{ g}\right)$
$$= 4200 \text{ cal}$$
$$\text{(calculator and}$$
$$\text{correct answer)}$$

Step 2. Solid copper at 1083°C to solid copper at 25°C

ht released = sp ht x mass x temperature change

$$= \left(\frac{0.093 \text{ cal}}{g\,^{\circ}C}\right) \times \left(100.0 \text{ g}\right) \times \left[(1083 - 25)\,^{\circ}C\right]$$

$= 9839.4$ cal (calculator answer)
$= 9800$ cal (correct answer)

Step 3. Add steps 1 and 2

$(4200 + 9800)$ cal $= 14000$ cal (calculator answer)
$= 14\overline{0}00$ cal (correct answer)

6. a) **Step 1.** Solid Al at 600°C to solid Al at 658°C

ht required = sp ht x mass x temperature change

$$= \left(\frac{.28 \text{ cal}}{g\,^{\circ}C}\right) \times \left(\overline{500} \text{ g}\right) \times \left[(658 - 600)\,^{\circ}C\right]$$

$= 8120$ cal (calculator answer)
$= 8100$ cal (correct answer)

Step 2. Solid Al at 658°C to liquid Al at 658°C

ht required = ht fusion x mass

$$= \left(76.8 \text{ cal/g}\right) \times \left(\overline{500} \text{ g}\right) = 38400 \text{ cal}$$

Step 3. Add steps 1 and 2

$(8100 + 38400)$ cal $= 46500$ cal (calculator and
correct answer)

b) **Step 1.** Solid Pb at 300°C to solid lead at 327°C

ht required = sp ht x mass x temperature change

$$= \left(\frac{.036 \text{ cal}}{g\,^{\circ}C}\right) \times \left(1.00 \text{ kg}\right) \times \left(\frac{10^3 \text{ g}}{1 \text{ kg}}\right)$$
$$\times \left[(327 - 300)\,^{\circ}C\right]$$

$= 972$ cal (calculator answer)
$= 970$ cal (correct answer)

Step 2. Solid Pb at 327°C to liquid Pb at 327°C

ht required = ht fusion x mass

$$= \left(5.86 \text{ cal/g}\right) \times \left(1.00 \text{ kg}\right) \times \left(10^3 \text{ g/kg}\right)$$

$$= 5860 \text{ cal (calculator and} \atop \text{correct answer)}$$

Step 3. Add steps 1 and 2

$$970 \text{ cal} + 5860 \text{ cal} = 6830 \text{ cal (calculator and} \atop \text{correct answer)}$$

c) **Step 1.** Solid gold at $21^{\circ}C$ to solid gold at $1064^{\circ}C$

ht. required = sp ht x mass x temperature change

$$= \left(\frac{.03 \text{ cal}}{g^{\circ}C}\right) \times \left(10.0 \text{ g}\right) \times \left[(1064 - 21)^{\circ}C\right]$$

$$= 312.9 \text{ cal (calculator answer)}$$
$$= 300 \text{ cal (correct answer)}$$

Step 2. Solid gold at $1064^{\circ}C$ to liquid gold at $1064^{\circ}C$

ht. required = ht fusion x mass $= \left(15.8 \text{ cal/g}\right) \times \left(10.0 \text{ g}\right)$

$$= 158 \text{ cal (calculator and} \atop \text{correct answer)}$$

Step 3. Add steps 1 and 2

$$300 \text{ cal} + 158 \text{ cal} = 458 \text{ cal (calculator answer)}$$
$$= 500 \text{ cal (correct answer)}$$

ANSWERS TO MULTIPLE CHOICE EXERCISES

1. d   10. c

2. b   11. c

3. c   12. b

4. b   13. b

5. c   14. a

6. a   15. a

7. c   16. b

8. c   17. a

9. c   18. b

CHAPTER 12
Gas Laws

REVIEW OF CHAPTER OBJECTIVES

1. <u>Understand the restriction on temperature-scale use required by the gas laws</u> (section 12.1).

   Three temperature scales were discussed in Chapter Three: Celsius, Fahrenheit and Kelvin. Only the Kelvin scale may be used in gas law calculations. If temperature is given in Celsius or Fahrenheit, it must be converted to Kelvin before it can be used in a gas law calculation.

   $$^{\circ}K = {}^{\circ}C + 273 \qquad\qquad {}^{\circ}C = \frac{5}{9}({}^{\circ}F - 32)$$

2. <u>State the definition of pressure, know the commonly used pressure units and be able to convert from one pressure unit to another</u> (section 12.1).

   Pressure is defined as force per unit area, that is,

   $$P = \frac{F}{A} \quad .$$

   A force of 120 pounds exerted on a surface area of one square inch (such as on a woman's narrow heel) produces a pressure of 120 lb/in^2. This is a much larger pressure than that produced by a force of 300 pounds exerted on a larger surface area of 6 square inches (such as a man's broad heel), which would be:

   $$P = \frac{F}{A} = \frac{300 \ lb}{6 \ in^2} = 50 \ lb/in^2$$

   The pressure exerted by a gas is due to the force exerted by gas molecules colliding with the walls of their container, or hitting any surface.

The interrelationships between the various pressure units are as follows:

1 atmosphere = 760 torr = 760 mm Hg = 29.92 in. Hg = 14.68 psi

The values 1 atm, 760 torr and 760 mm Hg are exact numbers since they arise from definitions. The values 29.92 in.Hg and 14.68 psi are measured numbers given to 4 significant figures.

Example: A laboratory barometer reads 752.5 mm Hg. What is this pressure in a) atmospheres, b) torr, c) inches of Hg, d) psi?

Solution:

a) 752.5 mm Hg = ? atm

$$752.5 \text{ mm Hg} \times \left(\frac{1 \text{ atm}}{760 \text{ mm Hg}}\right) = 0.99013158 \text{ atm}$$
(calcualtor answer)
$$= 0.9901 \text{ atm}$$
(correct answer)

b) 752.5 mm Hg = ? torr

$$752.5 \text{ mm Hg} \times \left(\frac{1 \text{ torr}}{1 \text{ mm Hg}}\right) = 752.5 \text{ torr}$$
(calculator and correct answer)

c) 752.5 mm Hg = ? in Hg

$$752.5 \text{ mm Hg} \times \left(\frac{1 \text{ atm}}{760 \text{ mm Hg}}\right) \times \left(\frac{29.92 \text{ in Hg}}{1 \text{ atm}}\right)$$
$$= 29.624737 \text{ in Hg (calculator answer)}$$
$$= 29.62 \text{ in Hg (correct answer)}$$

d) $$752.5 \text{ mm Hg} \times \left(\frac{1 \text{ atm}}{760 \text{ mm Hg}}\right) \times \left(\frac{14.68 \text{ psi}}{1 \text{ atm}}\right)$$
$$= 14.535132 \text{ psi (calculator answer)}$$
$$= 14.54 \text{ psi (correct answer)}$$

3. <u>Describe the principles of operation of barometers and manometers</u> (section 12.1).

The barometer is the standard device for measuring atmospheric pressure. A mercury barometer is made by filling a long glass tube, sealed at one end, with mercury and inverting the open end of the tube into a dish of mercury. The mercury in the tube falls until the pressure exerted by the mercury in the tube equals the pressure of the atmosphere on the mercury in the open dish (see Figure 12.1 on page 303 in the textbook).

A manometer is a device used to measure the pressure of a gas in a container. The manometer is a U-tube filled with mercury. One side of the U-tube is connected to the container in which the pressure is to be measured and the other side is connected to a region of known pressure. A difference in the heights of the mercury columns in the two arms of the manometer indicates the difference in pressure between the gas (in the container) and the atmosphere (see Figure 12.1 on page 305 in the textbook).

4. <u>State Boyle's Law and use it to solve problems involving changes in the condition of a gas</u> (section 12.1).

Boyle's law is a pressure-volume relationship. Boyle's law states that the volume of a sample of gas at constant temperature is inversely proportional to the pressure. When two quantities are inversely proportional, one increases as the other decreases. If pressure increases, volume decreases; if pressure decreases, volume increases.

Stated mathematically, Boyle's law states:

$$PV = k$$

If we start with a given sample of a gas with a volume = V_1 and a pressure = P_1, hold the temperature constant and change the pressure to P_2 (or the volume to V_2), then:

$$P_1V_1 = P_2V_2$$

When we know any three of the four quantities in this equation we can calculate the fourth one.

Example: At $25^{\circ}C$ and a pressure of 755 torr a sample of N_2 gas occupies a volume of 18.0 liters. What volume will this N_2 gas occupy if the pressure is increased to 825 torr at constant temperature?

Solution:

$$P_1 = 755 \text{ torr} \qquad P_2 = 825 \text{ torr}$$
$$V_1 = 18.0 \text{ liters} \qquad V_2 = ? \text{ liters}$$

$$P_1V_1 = P_2V_2 \qquad V_2 = \frac{V_1P_1}{P_2} = \frac{18.0 \text{ L} \times 755 \text{ torr}}{825 \text{ torr}}$$

$$= 16.472727 \text{ L (calculator answer)}$$

$$= 16.5 \text{ L (correct answer)}$$

Example: At $25^{\circ}C$, a sample of H_2 gas occupies a volume of 18.0 liters at a pressure of 755 torr. What pressure must be exerted on this H_2 gas, at constant temperature, to make its volume 25.0 liters?

Solution:

$$P_1 = 755 \text{ torr} \qquad P_2 = ? \text{ torr}$$

$$V_1 = 18.0 \text{ liters} \qquad V_2 = 25.0 \text{ liters}$$

$$P_1V_1 = P_2V_2 \qquad P_2 = \frac{P_1V_1}{V_2} = \frac{755 \text{ torr x } 18.0 \text{ L}}{25.0 \text{ L}}$$

$$= 543.6 \text{ torr (calculator answer)}$$
$$= 544 \text{ torr (correct answer)}$$

5. <u>State Charles's law and use it to solve problems involving changes in the conditions of a gas</u> (section 12.3).

Charles' law is a temperature-volume relationship. Charles' law states that the volume of a sample of a gas at constant pressure is directly proportional to its Kelvin temperature. When two quantities are directly proportional one increases when the other increases, and one decreases when the other decreases. If temperature increases, volume increases; if temperature decreases, volume decreases.

Stated mathematically, Charles' law is:

$$\frac{V}{T} = k$$

If we start with a given sample of a gas with a volume V_1 and a Kelvin temperature T_1, keep the pressure unchanged and change the Kelvin temperature to T_2 (or the volume to V_2), then:

$$\frac{V_1}{T_1} = \frac{V_2}{T_2}$$

With this equation we can solve for any one of the four quantities when the other three are known.

Example: A sample of Cl_2 gas occupies a volume of 458 mL at $10^{\circ}C$ and 1.11 atm pressure. What volume will this Cl_2 gas occupy at constant pressure if the temperature changes to $20^{\circ}C$?

Solution:

$$T_1 = 10^{\circ}C + 273 = 283 \text{ K} \qquad T_2 = 20^{\circ}C + 273 = 293 \text{ K}$$

$$V_1 = 458 \text{ mL} \qquad\qquad V_2 = ? \text{ mL}$$

$$\frac{V_1}{T_1} = \frac{V_2}{T_2} \qquad V_2 = \frac{V_1 T_2}{T_1} = \frac{458 \text{ mL} \times 293 \text{ K}}{283 \text{ K}}$$

$$= 474.18375 \text{ mL}$$
$$\text{(calculator answer)}$$
$$= 474 \text{ mL (correct answer)}$$

Example: A gas occupies a volume of 888 mL at $20^{\circ}C$ and 750 torr. If the pressure remains unchanged, at what temperature, in $^{\circ}C$, will the gas occupy a volume of 725 mL?

Solution:

$$V_1 = 888 \text{ mL} \qquad\qquad V_2 = 725 \text{ mL}$$

$$T_1 = 20^{\circ}C + 273 = 293 \text{ K} \qquad T_2 = ?^{\circ}C$$

$$\frac{V_1}{T_1} = \frac{V_2}{T_2} \qquad T_2 = \frac{T_1 \times V_2}{V_1} = \frac{293 \text{ K} \times 725 \text{ mL}}{888 \text{ mL}}$$

$$= 239.21734 \text{ K (calculator answer)}$$
$$= 239 \text{ K (correct answer)}$$

$$^{\circ}C = 239 \text{ K} - 273 = -34^{\circ}C$$

6. State Gay-Lussac's law, and use it to solve problems involving changes in the conditions of a gas (section 12.4).

Gay-Lussac's law is a temperature-pressure relationship. Gay-Lussac's law states that the pressure of a sample of a gas at constant volume is directly proportional to its Kelvin temperature. Thus, at constant volume, as temperature increases, pressure increases; as temperature decreases, pressure decreases.

Stated mathematically, Gay-Lussac's law is:

234   Gas Laws

$$\frac{P}{T} = k$$

If we start with a given sample of a gas at a pressure P_1 and a Kelvin temperature T_1, and keep the volume unchanged (as in a sealed glass bottle or a steel tank) while changing the temperature to T_2 (or the pressure to P_2), then:

$$\frac{P_1}{T_1} = \frac{P_2}{T_2}$$

With this equation in four unknowns, we can solve for any one given values for the other three.

Example: A gas exerts a pressure of 755 at 25°C. If the volume remains constant, what pressure will this gas exert at 50°C?

Solution:

According to Gay-Lussac's law, the pressure of a gas is directly proportional to its Kelvin temperature.

$P_1 = 755$ torr $P_2 = ?$ torr

$T_1 = 25°C + 273 = 298$ K $T_2 = 50°C + 273 = 323$ K

$$\frac{P_1}{T_1} = \frac{P_2}{T_2} \qquad P_2 = \frac{P_1 T_2}{T_1} = \frac{755 \text{ torr} \times 323 \text{ K}}{298 \text{ K}}$$

$$= 818.33893 \text{ torr (calculator answer)}$$

$$= 818 \text{ torr (correct answer)}$$

Example: 25.0 mL of a gas exerts a pressure of 725 torr at 18°C. When the volume is kept unchanged, at what temperature in °C will the pressure be 500 torr?

Solution:

$P_1 = 725$ torr $P_2 = 500$ torr

$T_1 = 18°C + 273 = 291$ K $T_2 = ?$ °C

$$\frac{P_1}{T_1} = \frac{P_2}{T_2} \qquad T_2 = \frac{T_1 \times P_2}{P_1} = \frac{291 \text{ K} \times 500 \text{ torr}}{725 \text{ torr}} = 201 \text{ K}$$

$$^\circ C = 201 \text{ K} - 273 = -72^\circ C$$

7. <u>Use the combined gas law to calculate, for a fixed</u>
    <u>amount of gas, the new value for a gas law variable</u>
    <u>brought about by changes in the other two gas law</u>
    <u>variables</u> (section 12.5).

    The combined gas law is a single expression obtained from
    mathematically combining Boyle's, Charles' and Gay-
    lussac's laws:

    $$\frac{P_1 V_1}{T_1} = \frac{P_2 V_2}{T_2}$$

    With the law, a change in any one of the three gas-law
    variables (P, V or T), brought about by changes in both
    of the other two variables, can be calculated. Each of
    the individual gas laws requires that one of the three
    variables be held constant.

    Example: A sample of O_2 gas occupies a volume of 952 mL
    at $25^\circ C$ and 762 torr. What volume, in mL, will
    this O_2 gas occupy at $10^\circ C$ and 745 torr?

Solution:

$$P_1 = 762 \text{ torr} \qquad\qquad P_2 = 745 \text{ torr}$$

$$V_1 = 952 \text{ mL} \qquad\qquad V_2 = ? \text{ mL}$$

$$T_1 = 25^\circ C + 273 = 298 \text{ K} \qquad T_2 = 10^\circ C + 273 = 283 \text{ K}$$

$$V_2 = V_1 \times \frac{T_2}{T_1} \times \frac{P_1}{P_2} = 952 \text{ mL} \times \left(\frac{283 \cancel{K}}{298 \cancel{K}}\right) \times \left(\frac{762 \cancel{\text{torr}}}{745 \cancel{\text{torr}}}\right)$$

$$= 924.71056 \text{ mL (calculator answer)}$$

$$= 925 \text{ mL (correct answer)}$$

    Example: At $15^\circ C$ and 1.10 atm pressure a sample of N_2
    gas occupies a volume of 75.5 liters. At what
    temperature, in $^\circ C$, will this N_2 gas occupy a
    volume of 85.5 liters and exert a pressure of
    570 torr?

Solution:

    Temperatures must be in Kelvin, and both pressures
    must have the same units.

$$V_1 = 75.5 \text{ liters} \qquad\qquad V_2 = 85.5 \text{ liters}$$

$$P_1 = 1.10 \text{ atm} \qquad\qquad P_2 = 570 \text{ torr} \times \frac{1 \text{ atm}}{760 \text{ torr}}$$

$$= 0.750 \text{ atm}$$

$$T_1 = 15^{\circ}C + 273 = 288 \text{ K} \qquad T_2 = ? \ ^{\circ}C$$

$$T_2 = T_1 \times \frac{P_2}{P_1} \times \frac{V_2}{V_1} = 288 \text{ K} \times \left(\frac{0.750 \ \cancel{atm}}{1.10 \ \cancel{atm}}\right) \times \left(\frac{85.5 \ \cancel{L}}{75.5 \ \cancel{L}}\right)$$

$$= 222.37206 \text{ K (calculator answer)}$$

$$= 222 \text{ K (correct answer)}$$

$$222 \text{ K} - 273 = -51^{\circ}C$$

8. <u>Define STP conditions</u> (section 12.6).

The volume of a gas changes considerably with changes in temperature and pressure (unlike liquids and solids, which undergo very little change in volume with temperature and pressure changes). Therefore, comparisons of gas volumes may be made only if the gases are at the same temperature and pressure. The temperature and pressure used for comparison purposes are:

Standard temperature = $0^{\circ}C$ = 273 K

and

Standard pressure = 1 atm = 760 torr

Standard temperature and standard pressure together are referred to as standard conditions, often called STP conditions.

Example: A sample of Cl_2 gas occupies a volume of 8.50 liters at $25^{\circ}C$ and 750 torr. What volume will this Cl_2 gas occupy at STP?

Solution:

STP = $0^{\circ}C$ and 760 torr

$$P_1 = 750 \text{ torr} \qquad\qquad P_2 = 760 \text{ torr}$$

$$V_1 = 8.50 \text{ liters} \qquad\qquad V_2 = ? \text{ liters}$$

$$T_1 = 25^{\circ}C + 273 = 298 \text{ K} \qquad T_2 = 0^{\circ}C + 273 = 273 \text{ K}$$

$$V_2 = V_1 \times \frac{P_1}{P_2} \times \frac{T_2}{T_1} = 8.50 \text{ L} \times \left(\frac{750 \text{ torr}}{760 \text{ torr}}\right) \times \left(\frac{273 \text{ K}}{298 \text{ K}}\right)$$

$$= 7.6844534 \text{ L (calculator answer)}$$

$$= 7.68 \text{ L (correct answer)}$$

9. <u>State Avogadro's law in words and mathematically</u>, and use <u>it in problem solving</u> (section 12.7).

Avogadro's law is a volume-quantity relationship. It states that equal volumes of different gases at the same temperature and pressure contain equal numbers of molecules.

Stated mathematically, with n = number of moles, Avogadro's law states:

$$\frac{V}{n} = k \qquad \text{or} \qquad \frac{V_1}{n_1} = \frac{V_2}{n_2}$$

Combining Avogadro's law and the combined gas las gives the expression:

$$\frac{P_1 V_1}{n_1 T_1} = \frac{P_2 V_2}{n_2 T_2}$$

This equation covers the situation where none of the four variables P, T, V and n, is constant.

Example: A 5.0 liter flask contains .208 moles of O_2 at a pressure of 1.02 atm and a temperature of 25°C. An additional .100 moles of O_2 is added, the volume remains unchanged and the temperature rises to 30°C. What is the new pressure exerted by the gas?

Solution:

Since the volume is unchanged, V_1 is equal to V_2 and volume will cancel out of the equation.

$$\frac{P_1 V_1}{n_1 T_1} = \frac{P_2 V_2}{n_2 T_2}$$

This gives the equation:

$$\frac{P_1}{n_1 T_1} = \frac{P_2}{n_2 T_2}$$

which is rearranged to isolate P_2 on the left side.

$$P_2 = P_1 \times \frac{n_2}{n_1} \times \frac{T_2}{T_1}$$

$P_1 = 1.02$ atm $P_2 = ?$ atm

$T_1 = 25^\circ C + 273 = 298$ K $T_2 = 30^\circ C + 273 = 303$ K

$n_1 = 0.208$ moles $n_2 = 0.208$ moles
                                          $+ 0.100$ moles
                                          $= 0.308$ moles

$P_2 = 1.02$ atm $\times \dfrac{0.308 \ \text{moles}}{0.208 \ \text{moles}} \times \dfrac{303 \ \text{K}}{298 \ \text{K}}$

    $= 1.5357266$ atm (calculator answer)

    $= 1.54$ atm (correct answer)

10. <u>Define molar volume and use this relationship in problem-solving situations</u> (section 12.8).

The molar volume of a gas is the volume occupied by one mole of the gas at STP conditions. The molar volume of any gas is 22.4 liters. From this concept we get the following conversion factors at STP conditions:

1 mole gas = 22.4 liters

so, $\dfrac{1 \text{ mole gas}}{22.4 \text{ L gas}}$ or $\dfrac{22.4 \text{ L gas}}{1 \text{ mole gas}}$

A most common type of problem involving these conversion factors is one where the volume of a gas at STP is known and you are asked to calculate from it either the moles, grams or particles of gas present, or vice versa. The relationships needed in working such problems are summarized in the following diagram (which is also given as Figure 12-6 in the textbook).

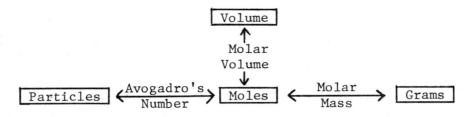

Example: What is the volume at STP, in liters, occupied by 5.50 grams of Cl_2 gas?

Solution:

5.50 g Cl_2 gas = ? liters Cl_2 at STP

This is a grams-to-volume problem. The pathway, determined using the previous diagram, is:

g Cl_2 ⟶ moles Cl_2 ⟶ liters Cl_2 at STP

$$5.50 \text{ g } Cl_2 \times \left(\frac{1 \text{ mole } Cl_2}{71.0 \text{ g } Cl_2}\right) \times \left(\frac{22.4 \text{ L } Cl_2}{1 \text{ mole } Cl_2}\right)$$

$$= 1.7352113 \text{ L } Cl_2 \text{ (calculator answer)}$$

$$= 1.74 \text{ L } Cl_2 \text{ (correct answer)}$$

Whenever the formula of a gas is known, its density at STP can be calculated from its molar mass and molar volume.

Example: Calculate the density at STP conditions, in grams/liter, of SO_3 gas.

Solution:

1 mole SO_3 gas = 22.4 liters (molar volume)

1 mole SO_3 gas = 80.1 grams (molar mass)

$$d = \left(\frac{80.1 \text{ g } SO_3}{1 \text{ mole } SO_3}\right) \times \left(\frac{1 \text{ mole } SO_3}{22.4 \text{ L } SO_3}\right)$$

$$= 3.5758929 \text{ g/L (calculator answer)}$$

$$= 3.58 \text{ g/L (correct answer)}$$

11. <u>Give the units and value of the ideal gas constant, R, for commonly encountered unit combinations</u> (section 12.9).

The ideal gas law is usually written as PV = nRT. The value R in this equation is called the ideal gas law constant. Its value can be found by substituting values for the other 4 variables in the equation and solving for R.

At STP, P = 1 atm, V = 22.4 L, T = 273 K, and n = 1 mole

Substituting these values in the equation PV = nRT and solving for R gives:

$$R = \frac{PV}{nT} = \frac{1 \text{ atm} \times 22.4 \text{ L}}{1 \text{ mole} \times 273 \text{ K}} = 0.08205128 \frac{\text{atm-L}}{\text{mole-K}} \text{ (calculator answer)}$$

$$= 0.0821 \frac{\text{atm-L}}{\text{mole-K}} \text{ (correct answer)}$$

If standard pressure is used as 760 torr, then:

$$R = \frac{PV}{nT} = \frac{760 \text{ torr} \times 22.4 \text{ L}}{1 \text{ mole} \times 273 \text{ K}} = 62.358974 \frac{\text{torr-L}}{\text{mole-K}} \text{ (calculator answer)}$$

$$= 62.4 \frac{\text{torr-L}}{\text{mole-K}} \text{ (correct answer)}$$

For other pressure or volume or temperature units convert them to the units in these known R values.

12. <u>Use the ideal gas law to calculate a value for P, T, V or n given values for the other three</u> (section 12.9).

The ideal gas law, PV = nRT, is used in calculations when <u>one</u> set of conditions is given with one missing variable (either P, V, T or n). It is not directly applicable when <u>two</u> sets of conditions are given with one missing variable. In this latter situation it is easier to use the combined gas law.

Example: Calculate the temperature, in $^\circ$C, at which 2.50 moles of H_2 occupies a volume of 55.0 liters at 1.35 atm.

Solution:

P = 1.35 atm   n = 2.50 moles

V = 55.0 liters $R = 0.0821 \frac{\text{atm-L}}{\text{mole-}^\circ\text{K}}$

T = ? $^\circ$C

PV = nRT

$$T = \frac{PV}{nR} = \frac{1.35 \text{ atm} \times 55.0 \text{ L}}{2.50 \text{ moles} \times .0821 \frac{\text{atm-L}}{\text{mole-}^{\circ}K}}$$

$$= 361.75396 \text{ K (calculator answer)}$$

$$= 362 \text{ K (correct answer)}$$

$$362 \text{ K} - 273 = 89^{\circ}C$$

Example: What volume, in liters, is occupied by 0.126 moles of H_2 gas at 752 torr and $22^{\circ}C$?

Solution:

P = 752 torr   n = 0.126 moles

$T = 22^{\circ}C + 273 = 295^{\circ}K$ $R = 62.4 \frac{\text{torr-L}}{\text{mole-}^{\circ}K}$

V = ?

PV = nRT

$$V = \frac{nRT}{P} = \frac{.126 \text{ moles} \times 62.4 \frac{\text{torr-L}}{\text{mole-}K} \times 295 \text{ K}}{762 \text{ torr}}$$

$$= 3.0438425 \text{ L (calculator answer)}$$

$$= 3.04 \text{ L (correct answer)}$$

13. Use the ideal gas law in modified form to calculate the mass of a gas, its molecular weight or its density from appropriate data (section 12.10).

Useful modified forms of the ideal gas law, whose derivations are given in the textbook on pages 328 and 329, are:

(1) $g = \frac{PV(MW)}{RT}$

(2) $MW = \frac{gRT}{PV}$

(3) $d = \frac{P(MW)}{RT}$

Equation 1 is used to calculate the mass, in grams of a gas; equation 2 finds use in molecular weight calculations; equation 3 enables one to calculate directly the density of a gas of known molecular weight.

Example: How many grams of CO occupy a volume of 0.0250 L at 10°C and 2.50 atm?

Solution:

P = 2.50 atm

V = 0.0250 L

MW = 28.0 g/mole

$R = 0.0821 \dfrac{atm-L}{mole-K}$

$T = 10^\circ C + 273 = 283\ K$

Using the ideal gas law in the form $g = \dfrac{P \times V \times MW}{R \times T}$

and substituting in known quantities gives

$$g = \dfrac{2.50\ \cancel{atm} \times .0250\ \cancel{L} \times 28.0\ g/\cancel{mole}}{.0821\ \dfrac{\cancel{atm-L}}{\cancel{mole-K}} \times 183\cancel{K}}$$

$$= 0.75319678\ g\ (calculator\ answer)$$

$$= .0753\ g\ (correct\ answer)$$

Example: Find the molecular weight of a gas if 50.6 grams of this gas occupy a volume of 17.1 liters at 780 torr and 27°C.

Solution:

P = 780 torr

$T = 27^\circ C + 273 = 300\ K$

g = 50.6 g

V = 17.1 liters

$R = 62.4 \dfrac{torr-L}{mole-K}$

Using the ideal gas equation in the form

$$MW = \dfrac{g \times R \times T}{P \times V}$$

and substituting in known quantities gives

$$MW = \dfrac{50.6\ g \times 62.4\ \dfrac{\cancel{torr-L}}{mole-\cancel{K}} \times 300\ \cancel{K}}{780\ \cancel{torr} \times 17.1\ \cancel{L}}$$

$$= 71.017544\ g/mole\ (calculator\ answer)$$

$$= 71.0\ g/mole\ (correct\ answer)$$

Example: Calculate the density, in grams/liter, of O_2 gas at $30°C$ and 1.21 atm.

Solution:

$P = 1.21$ atm $\qquad R = .0821 \dfrac{\text{atm-L}}{\text{mole-K}}$

$MW = 32.0$ g/mole

$\qquad\qquad\qquad T = 30°C + 273 = 303$ K

Using the "density" form of the ideal gas law gives:

$$d = \dfrac{g}{V} = \dfrac{P \times MW}{R \times T}$$

$$= \dfrac{1.21 \ \text{atm} \times 32.0 \ \text{g/mole}}{.0821 \ \dfrac{\text{atm-L}}{\text{mole-K}} \times 303 \ \text{K}} = 1.5565016 \ \text{g/L}$$

(calculator answer)

$$= 1.56 \ \text{g/L}$$
(correct answer)

14. <u>Understand the relationship at constant temperature and pressure between the volumes of different gases consumed or produced in a chemical reaction</u> (section 12.11).

Gases are often reactants or products in chemical reactions, and it is easier to measure volumes of gases than weights of gases. The volumes of different gases involved in a chemical reaction, when measured at the same pressure and temperature, have the same ratio as the coefficients in the balanced equation. This coefficient-volume relationship follows directly from Avogadro's law, which says that moles and volumes of a gas are directly proportional at the same temperature and pressure.

In calculations involving chemical reactions, where two or more gases are participants, this volume interpretation of equation coefficients may be used to generate conversion factors useful in problem solving. For example, from the balanced equation:

$$3H_{2(g)} + N_{2(g)} \longrightarrow 2NH_{3(g)}$$

six conversion factors may be derived:

$$\dfrac{3 \text{ volumes } H_2}{1 \text{ volume } N_2} \qquad\qquad \dfrac{1 \text{ volume } N_2}{3 \text{ volumes } H_2}$$

$$\frac{3 \text{ volumes } H_2}{2 \text{ volumes } NH_3} \qquad \frac{2 \text{ volumes } NH_3}{3 \text{ volumes } H_2}$$

$$\frac{1 \text{ volume } N_2}{2 \text{ volumes } NH_3} \qquad \frac{2 \text{ volumes } NH_3}{1 \text{ volume } N_2}$$

15. <u>Calculate the volume or mass of a gas, under a given set of conditions, that is required as a reactant or formed as a product in a chemical reaction</u> (section 12.11).

Mass-volume and volume-mass calculations are carried out with procedures which are just a simple extension of those needed for grams-to-moles-to-particles problems. The additional relationship needed is the molar volume relationship.

1 mole of a gas (at STP) = 22.4 liters of gas

or the ideal gas law.

Figure 12-7 in your textbook (page 333) summarizes all of the needed relationships.

Example: Hydrogen and oxygen react to **pro**duce water as shown by the following equation:

$$2H_{2(g)} + O_{2(g)} \longrightarrow 2H_2O_{(g)}$$

What mass of water is produced when 17.1 liters of O_2 at $20°C$ and 720 torr reacts with the needed amount of H_2?

Solution:

We first find the number of moles of O_2 using the ideal gas law.

$$n_{O_2} = \frac{PV}{RT}$$

$$= \frac{720 \text{ torr} \times 17.1 \text{ L}}{62.4 \frac{\text{torr} \cdot \text{L}}{\text{mole} \cdot \text{K}} \times 293 \text{ K}} = 0.67340509 \text{ moles } O_2$$
(calculator answer)

$$= 0.673 \text{ moles } O_2$$
((correct answer)

We then go from moles of O_2 to grams H_2O with a

dimensional analysis set-up

$$0.673 \text{ moles } O_2 \text{ x} \left(\frac{2 \text{ moles } H_2O}{1 \text{ mole } O_2} \right) \text{x} \left(\frac{18.0 \text{ g } H_2O}{1 \text{ mole } H_2O} \right)$$

$$= 24.228 \text{ g } H_2O \text{ (calculator answer)}$$

$$= 24.3 \text{ g } H_2O \text{ (correct answer)}$$

Alternatively, we could have obtained the moles of O_2 (the first step) by using the combined gas law to change the given volume from nonstandard to STP conditions and then using the molar volume relationship

$$17.1 \text{ L } O_2 \text{ x} \left(\frac{273 \text{ K}}{293 \text{ K}} \right) \text{x} \left(\frac{720 \text{ torr}}{760 \text{ torr}} \right) \text{x} \left(\frac{1 \text{ mole } O_2}{22.4 \text{ liter}} \right)$$

$$= 0.67384812 \text{ moles } O_2 \text{ (calculator answer)}$$

$$= 0.674 \text{ moles } O_2 \text{ (correct answer)}$$

The first two conversion factors in this set-up come from the equation

$$V_2 = V_1 \text{ x } \frac{T_2}{T_1} \text{ x } \frac{P_1}{P_2}$$

16. <u>Use Dalton's law of partial pressures to calculate the partial or total pressures from appropriate data on mixtures of gases and also to correct for the effects of water vapor pressure in gases collected by water displacement</u> (section 12.12).

The pressure of a mixture of gases equals the sum of the pressures of each of the individual gases.

$$P_{air} = P_{O_2} + P_{N_2} + \text{etc.}$$

A widely used example of Dalton's law of partial pressure applies to gases collected over water. If H_2 gas is bubbled through water, the sample of gas collected contains both H_2 and H_2O vapor. If this sample of gas exerts a pressure of 750 torr, this pressure is due to the pressures of both H_2 and H_2O vapor.

The pressure of a gas collected over water is given by the equation

$$P_{gas} = P_{atm} - P_{H_2O}$$

where P_{H_2O} is the vapor pressure of the water. A table of water vapor pressures at various temperatures is given on page 339 in the textbook.

Example: The pressure of a gas, H_2, collected over water is 750 torr, and the pressure of the H_2 alone is 730 torr. What is the partial pressure of H_2O vapor?

Solution:

$$P_T = P_{H_2} + P_{H_2O} \quad \text{and} \quad P_{H_2O} = P_T - P_{H_2}$$

$$= 750 \text{ torr} - 730 \text{ torr} = 20 \text{ torr}$$

The partial pressure of H_2O vapor depends only on the temperature of the water. The vapor pressure of water increases with temperature.

Example: What is the partial pressure of H_2 collected over water at $21^\circ C$ when atmospheric pressure is 755 torr?

Solution:

The vapor pressure table on page 339 of the textbook indicates that the partial pressure of water vapor at $21^\circ C$ is 18.7 torr.

$$P_{H_2} = P_{atm} - P_{H_2O}$$

$$= (755 - 18.7) \text{ torr}$$

$$= 736.3 \text{ torr (calculator answer)}$$

$$= 736 \text{ torr (correct answer)}$$

PROBLEM SET

1. Express each pressure in the requested unit.

   a) 720 torr = ? atm
   b) 35.5 in Hg = ? torr
   c) 13.9 psi = ? atm
   d) 5.25 atm = ? torr

2. A sample of CH_4 gas occupies a volume of 12.8 liters at 600 torr and $22^{\circ}C$. What volume will this gas occupy at 700 torr and the same temperature?

3. A sample of He gas in an expandable balloon occupies a volume of 4.00 liters at $10^{\circ}C$ and 800 torr. What pressure, in torr, must be exerted on the He gas in the balloon at constant temperature to change the volume to 1.75 liters?

4. A sample of CO_2 occupies a volume of 1.23 liters at $17^{\circ}C$ and 750 torr. What volume will the CO_2 occupy at $-10^{\circ}C$ if the pressure remains unchanged?

5. A sample of Ne gas occupies a volume of 50.0 liters at $75^{\circ}C$ and 580 torr. At what temperature, in $^{\circ}C$, will the volume of the gas be 25.0 liters if the pressure remains the same?

6. A 10.0 liter steel tank contains He gas at $22^{\circ}C$ and 2.25 atm. What will be the pressure of the He gas when the temperature changes to $80^{\circ}C$?

7. A sample of acetylene gas, C_2H_2, occupies a volume of 285 mL at $80^{\circ}C$ and 800 torr. What volume will this C_2H_2 occupy at 10° C and 750 torr?

8. A balloon which contains He gas has a volume of 30.0 liters high above the surface of the earth, where the temperature is $-20.0^{\circ}C$ and the pressure is 380 torr. What volume did the He occupy at the surface of the earth when the temperature was $25.0^{\circ}C$ and the pressure was 748 torr?

9. A sample of gas occupies a volume of 859 mL at $29^{\circ}C$ and 763 torr. Find the volume of the gas at STP conditions.

10. 1.73 moles of CO occupy a volume of 42.0 liters at a given temperature and pressure.

    a) What volume will 2.00 moles of CO occupy at the same temperature and pressure?

    b) What volume will 3.00 moles of He occupy at the same temperature and pressure?

11. Calculate the volume in liters of 50.0 g of each of the following gases at STP conditions:

    a) Ne   b) N_2 c) NO_2

12. Calculate the density at STP conditions of the following gases:

a) Ar   b) CH_4 c) C_2H_6

13. Use the ideal gas law to calculate the volume of:

a) 2.30 moles of O_2 at $30°C$ and 1.20 atm

b) 4.73 moles of N_2 at $10°C$ and 745 torr

c) 2.73 grams of CO at $-10°C$ and 1.10 atm

14. Calculate the molecular weight of a gas if 0.181 grams of the gas occupies a volume of 250 mL at $10°C$ and 800 torr.

15. Calculate the molecular weight of a gas which has a density of 1.34 g/L at STP conditions.

16. In the formation of oxygen gas by decomposition of $KClO_3$ according to the equation:

$$2KClO_3 \longrightarrow 2KCl + 3 O_2$$

how many grams of $KClO_3$ must be decomposed to produce 254 mL of O_2 gas at $24°C$ and 752 torr?

17. At $28°C$ and 748 torr, 225 mL of H_2 gas is collected over water. What volume will the H_2 gas occupy, dry, at STP?

MULTIPLE CHOICE EXERCISES

1. When using gas laws the temperature of a gas must always be specified on which temperature scale?

a) Celsius
b) Fahrenheit
c) Kelvin
d) Celvinheit

2. All of the following pressures are equivalent except for one. The exception is:

a) 670 torr   c) 14.7 psi
b) 1 atm   d) 760 mm Hg

3. Which of the following statements is false?

   a) air at 1 atm pressure will support a column of Hg 760 mm high
   b) the atmospheric pressure at sea level is always 760 torr
   c) a barometer is a standard device for measuring atmospheric pressure
   d) pressures expressed in torr and mm Hg will always be numerically equal

4. Indicate the missing words in the following statement of Boyle's law: "At _____ temperature the volume of a sample gas is _____ proportional to the pressure applied to the gas."

   a) increased and directly
   b) decreased and inversely
   c) constant and inversely
   d) constant and directly

5. Six liters of oxygen gas has a pressure of 500 torr. If the volume of this gas is decreased to 3 liters at constant temperature, the new pressure will be:

   a) 250 torr
   b) 500 torr
   c) 750 torr
   d) 1000 torr

6. A mathematical statement of Charles' law is:

   a) $V_1 + T_1 = V_2 + T_2$

   b) $V_1 T_1 = V_2 T_2$

   c) $\dfrac{V_1}{T_1} = \dfrac{V_2}{T_2}$

   d) $\dfrac{V_1}{T_2} = \dfrac{V_2}{T_1}$

7. According to Charles' law if the Kelvin temperature of a gas sample is doubled the gas sample:

   a) volume decreases by a factor of 2
   b) volume is doubled
   c) pressure decreases by a factor of 2
   d) pressure is doubled

8. Gay-Lussac's law involves which of the following:

   a) an indirect proportion
   b) a constant volume
   c) a constant pressure
   d) a constant temperature

9. According to the combined gas law, for a fixed amount of gas, volume and pressure:

   a) are inversely proportional
   b) are directly proportional
   c) have no relationship to each other
   d) are sometimes inversely and sometimes directly proportional

10. A sample of 6 liters of nitrogen gas has a pressure of 200 torr at 300 K. If the temperature is increased to 600 K and the pressure is decreased to 100 torr the new volume will be:

    a) 1.5 liters
    b) 3 liters
    c) 6 liters
    d) 24 liters

11. Which of the following is not STP conditions:

    a) $0^{\circ}C$, 760 torr
    b) 273 K, 760 mm Hg
    c) $0^{\circ}C$, 1 atm
    d) 0 K, 1 atm

12. Which of the following laws would be most useful in solving the problem "A 2.00 mole quantity of a gas has a volume of 2.00 L at a certain temperature and pressure. What is the volume of a 1.50 mole quantity of the same gas at the same conditions?"

    a) Gay-Lussac's law
    b) Avogadro's law
    c) Boyle's law
    d) Charles' law

13. At STP, the molar volume of any gas is:

    a) a volume in mL numerically equal to its molecular weight
    b) a volume in mL numerically equal to its density
    c) 22.4 mL
    d) 22.4 L

14. Which of the following is true of 8.0 g of He at STP?

    a) it has a pressure of 2 atm
    b) it has a volume of 22.4 L
    c) it is at a temperature of $27^{\circ}C$
    d) it has a volume of 44.8 L

15. The constant R in the ideal gas equation can be evaluated in terms of many different sets of units. Which of the following is a correct set of units?

    a) mole-K/torr-mL
    b) atm-mL/mole K
    c) torr-moles/L-K
    d) atm-K/moles-L

16. Which of the following is a correct expression for the density of a gas?

    a) PV/(MW)T
    b) (MW)P/RT
    c) (MW)RT/P
    d) PT/(MW)RV

17. According to the equation $H_{2(g)} + I_{2(g)} \longrightarrow 2HI_{(g)}$, which of the following statements is not true at constant temperature and pressure?

    a) the volume of H_2 needed is the same as the volume of I_2
    b) the volume of HI produced will be twice that of the H_2 used
    c) the volume of H_2 will be less than that of I_2 because H_2 is lighter than I_2
    d) the volume of HI produced is equal to the combined volumes of H_2 and I_2 consumed

18. Of the following gases, the one with the greatest density at STP is:

    a) CH_4
    b) NH_3
    c) Ne
    d) H_2

19. The sum of the partial pressures of all the gases in a gaseous mixture:

    a) is greater than the total pressure of the mixture
    b) is less than the total pressure of the mixture
    c) is equal to the total pressure of the mixture
    d) has no relationship to the total pressure of the mixture

20. A sample of O_2 gas is collected over water at $23^{\circ}C$ at a barometric pressure of 751 torr. The vapor pressure of water at $23^{\circ}C$ is 21 torr. The partial pressure of the O_2 gas in the sample collected is:

    a) 21 torr
    b) 728 torr
    c) 751 torr
    d) 774 torr

21. If 5.00 grams of a gas occupy a volume of 2.00 liters at STP conditions,

    a) the density of the gas at STP is 3.50 grams/liter
    b) 1.00 grams of this gas at STP would occupy a volume of 300 mL
    c) the molecular weight of the gas is 44.0 g/mole
    d) 10.0 grams of this gas would occupy a volume of 4.50 liters at $10^{\circ}C$ and 700 torr

SOLUTIONS TO PROBLEMS SET

1. a) $720 \text{ torr} \times \left(\dfrac{1 \text{ atm}}{760 \text{ torr}} \right) = 0.94736842$ atm
(calculator answer)
$= 0.947$ atm (correct answer)

   b) $35.5 \text{ in Hg} \times \left(\dfrac{1 \text{ atm}}{29.9 \text{ in Hg}} \right) \times \left(\dfrac{760 \text{ torr}}{1 \text{ atm}} \right) = 902.34114$ torr
(calculator answer)
$= 902$ torr
(correct answer)

   c) $13.9 \text{ psi} \times \left(\dfrac{1 \text{ atm}}{14.7 \text{ psi}} \right) = 0.94557823$ atm
(calculator answer)
$= 0.946$ atm (correct answer)

   d) $5.25 \text{ atm} \times \left(\dfrac{760 \text{ torr}}{1 \text{ atm}} \right) = 3990$ torr
(calculator and correct answer)

2. This is a Boyle's law problem: $P_1V_1 = P_2V_2$

$P_1 = \overline{6}00$ torr $P_2 = \overline{7}00$ torr

$V_1 = 12.8$ liters $V_2 = ?$ liters

$V_2 = \dfrac{V_1 \times P_1}{P_2} = 12.8 \text{ L} \times \dfrac{\overline{6}00 \text{ torr}}{\overline{7}00 \text{ torr}} = 10.971429$ L
(calculator answer)
= 11.0 L
(correct answer)

3. Boyle's law states that: $P_1V_1 = P_2V_2$

$P_1 = \overline{8}00$ torr $P_2 = \ell$ torr

$V_1 = 4.00$ liters $V_2 = 1.75$ liters

$P_2 = \dfrac{P_1V_1}{V_2} = \dfrac{\overline{8}00 \text{ torr} \times 4.00 \not{L}}{1.75 \not{L}} = 1828.5714$ torr
(calculator answer)
= 1830 torr
(correct answer)

4. This is a Charles' law problem: $\dfrac{V_1}{T_1} = \dfrac{V_2}{T_2}$

$V_1 = 1.23$ liters $V_2 = ?$ liters

$T_1 = 17^\circ C + 273 = 290$ K $T_2 = \overline{1}0^\circ C + 273 = 263$ K

$V_2 = \dfrac{V_1 T_2}{T_1} = \dfrac{1.23 \text{ L} \times 263 \not{K}}{290 \not{K}} = 1.1154828$ L
(calculator answer)
= 1.12 L (correct answer)

5. Charles' law states: $\dfrac{V_1}{T_1} = \dfrac{V_2}{T_2}$

$V_1 = 50.0$ liters $V_2 = 25.0$ liters

$T_1 = 75^\circ C + 273 = 348$ K $T_2 = ? \ ^\circ C$

$T_2 = \dfrac{T_1 V_2}{V_1} = \dfrac{348 \text{ K} \times 25.0 \not{L}}{50.0 \not{L}} = 174$ K (calculator and
correct answer)

$174 \text{ K} - 273 = -99^\circ C$

6. Gay-Lussac's law states: $\dfrac{P_1}{T_1} = \dfrac{P_2}{T_2}$

P_1 = 2.25 atm $\qquad\qquad$ P_2 = ? atm

T_1 = 22°C + 273 = 295 K $\qquad$ T_2 = 80°C + 273 = 353 K

$P_2 = \dfrac{P_1 T_2}{T_1}$ = $\dfrac{2.25 \text{ atm} \times 353 \cancel{K}}{295 \cancel{K}}$ = 2.6923729 atm (calculator answer)

$\qquad\qquad\qquad\qquad\qquad\qquad\qquad\qquad$ = 2.69 atm (correct answer)

7. The combined gas law states: $\dfrac{P_1 V_1}{T_1} = \dfrac{P_2 V_2}{T_2}$

P_1 = $80\overline{0}$ torr $\qquad\qquad$ P_2 = $75\overline{0}$ torr

V_1 = 285 mL $\qquad\qquad\qquad$ V_2 = ? mL

T_1 = 80°C + 273 = 353 K $\qquad$ T_2 = $1\overline{0}^{\circ}$C + 273 = 283 K

V_2 = 285 mL $\times$ $\dfrac{80\overline{0} \cancel{\text{torr}}}{75\overline{0} \cancel{\text{torr}}}$ $\times$ $\dfrac{283 \cancel{K}}{353 \cancel{K}}$ = 243.71671 mL (calculator answer)

$\qquad\qquad\qquad\qquad\qquad\qquad\qquad\qquad$ = 244 mL (correct answer)

8. Using the combined gas law, $\dfrac{P_1 V_1}{T_1} = \dfrac{P_2 V_2}{T_2}$, we get:

P_1 = $38\overline{0}$ torr $\qquad\qquad$ P_2 = 748 torr

V_1 = 30.0 liters $\qquad\qquad$ V_2 = ? liters

T_1 = -20°C + 273 = 253 K $\qquad$ T_2 = 25°C + 273 = 298 K

V_2 = 30.0 L $\times$ $\dfrac{298 \cancel{K}}{253 \cancel{K}}$ $\times$ $\dfrac{38\overline{0} \cancel{\text{torr}}}{748 \cancel{\text{torr}}}$ = 17.951428 L (calculator answer)

$\qquad\qquad\qquad\qquad\qquad\qquad\qquad\qquad$ = 18.0 L (correct answer)

9. Using the combined gas law, $\dfrac{P_1 V_1}{T_1} = \dfrac{P_2 V_2}{T_2}$, we get:

P_1 = 763 torr $\qquad\qquad$ P_2 = 760 torr

V_1 = 859 mL $\qquad\qquad\qquad$ V_2 = ? mL

T_1 = 29°C + 273 = 302 K $\qquad$ T_2 = 0°C + 273 = 273 K

$$V_2 = 859 \text{ mL } x\left(\frac{273 \cancel{K}}{302 \cancel{K}}\right)x\left(\frac{763 \cancel{torr}}{760 \cancel{torr}}\right) = 779.57843 \text{ mL}$$

= (calculator answer)

= 780 mL (correct answer)

10. Avogadro's law states $\dfrac{V_1}{n_1} = \dfrac{V_2}{n_2}$

a) $V_1 = 42.0 \text{ L}$ $\qquad$ $V_2 = ? \text{ L}$

$n_1 = 1.73 \text{ moles}$ $\qquad$ $n_2 = 2.00 \text{ moles}$

$$V_2 = 42.0 \text{ L } x\left(\frac{2.00 \cancel{moles}}{1.73 \cancel{moles}}\right) = 48.554913 \text{ L}$$

= (calculator answer)

= 48.6 L (correct answer)

b) as in part a), $\dfrac{V_1}{n_1} = \dfrac{V_2}{n_2}$

even though the gases are different, the volume is directly proportional to the number of moles.

$V_1 = 42.0 \text{ L}$ $\qquad$ $V_2 = ? \text{ L}$

$n_1 = 1.73 \text{ moles (CO)}$ $\qquad$ $n_2 = 3.00 \text{ moles (He)}$

$$V_2 = 42.0 \text{ L } x\left(\frac{3.00 \cancel{moles}}{1.73 \cancel{moles}}\right) = 72.83237 \text{ L}$$

= (calculator answer)

= 72.8 L (correct answer)

11. One mole of any gas at STP occupies a volume of 22.4 liters.

a) $50.0 \text{ g } \cancel{Ne} \text{ } x\left(\dfrac{1 \text{ mole } \cancel{Ne}}{20.2 \text{ g } \cancel{Ne}}\right)x\left(\dfrac{22.4 \text{ L}}{1 \text{ mole } \cancel{Ne}}\right) = 55.445545$

= (calculator answer)

= 55.4 L

(correct answer)

b) $50.0 \text{ g } \cancel{N}_2 \text{ } x\left(\dfrac{1 \text{ mole } \cancel{N}_2}{28.0 \text{ g } \cancel{N}_2}\right)x\left(\dfrac{22.4 \text{ L}}{1 \text{ mole } \cancel{N}_2}\right) = 40 \text{ L}$

= (calculator answer)

= 40.0 L

(correct answer)

c) $50.0 \text{ g } \cancel{NO}_2 \text{ } x\left(\dfrac{1 \text{ mole } \cancel{NO}_2}{46.0 \text{ g } \cancel{NO}_2}\right)x\left(\dfrac{22.4 \text{ L}}{1 \text{ mole } \cancel{NO}_2}\right)$

= 24.347826 L (calculator answer)

= 24.3 L (correct answer)

12. Using the molar mass and molar volume relationships and the fact that density is mass divided by volume we get:

a) $d_{Ar} = \left(\dfrac{39.9 \text{ g}}{1 \text{ mole}}\right) \times \left(\dfrac{1 \text{ mole}}{22.4 \text{ L}}\right) = 1.78125$ g/L (calculator answer)
$= 1.78$ g/L (correct answer)

b) $d_{CH_4} = \left(\dfrac{16.0 \text{ g}}{1 \text{ mole}}\right) \times \left(\dfrac{1 \text{ mole}}{22.4 \text{ L}}\right) = 0.71428571$ g/L
(calculator answer)
$= 0.714$ g/L (correct answer)

c) $d_{C_2H_6} = \left(\dfrac{30.1 \text{ g}}{1 \text{ mole}}\right) \times \left(\dfrac{1 \text{ mole}}{22.4 \text{ L}}\right) = 1.34375$ g/L
(calculator answer)
$= 1.34$ g/L (correct answer)

13. Using the ideal gas law we get:

a) $V = \dfrac{nRT}{P} = \dfrac{2.30 \text{ moles} \times .0821 \dfrac{\text{atm-L}}{\text{mole-K}} \times 303 \text{ K}}{1.20 \text{ atm}}$

$= 47.679575$ L (calculator answer)
$= 47.7$ L (correct answer)

b) $V = \dfrac{nRT}{P} = \dfrac{4.73 \text{ moles} \times 62.4 \dfrac{\text{torr-L}}{\text{mole-K}} \times 283 \text{ K}}{745 \text{ torr}}$

$= 112.11814$ L (calculator answer)
$= 112$ L (correct answer)

c) $V = \dfrac{gRT}{MW \, P} = \dfrac{2.73 \text{ g CO} \times .0821 \dfrac{\text{atm-L}}{\text{mole-K}} \times 263 \text{ K}}{28.0 \dfrac{\text{g CO}}{\text{mole}} \times 1.10 \text{ atm}}$

$= 1.913863$ L (calculator ans) $= 1.91$ L (correct ans)

14. Using the "molecular weight" form of the ideal gas law, $MW = \dfrac{gRT}{PV}$, we get:

$MW = \dfrac{0.181 \text{ g} \times 62.4 \dfrac{\text{torr-L}}{\text{mole-K}} \times 283 \text{ K}}{800 \text{ torr} \times .250 \text{ L}}$

$= 15.981576$ g/mole (calculator answer)
$= 16.0$ g/mole (correct answer)

15. $\left(\dfrac{1.34 \text{ g}}{1 \text{ liter}}\right) \times \left(\dfrac{22.4 \text{ liters}}{1 \text{ mole}}\right) = 30.016$ g/mole (calculator answer)
$= 30.0$ g/mole (correct answer)

16. $PV = nRT$, and $n = \dfrac{PV}{RT}$

$V = 254 \; mL \times \left(\dfrac{1 \; liter}{1000 \; mL}\right) = 0.254 \; liters$

$moles \; O_2 = \dfrac{PV}{RT} = \dfrac{752 \; torr \times 0.254 \; L}{62.4 \; \dfrac{torr \cdot L}{mole \cdot K} \times 297 \; K} = 0.01030648 \; moles \; O_2$

(calculator answer)

$= 0.0103 \; mole \; O_2$

(correct answer)

$0.0103 \; moles \; O_2 \times \left(\dfrac{2 \; moles \; KClO_3}{3 \; moles \; O_2}\right) \times \left(\dfrac{123 \; g \; KClO_3}{1 \; mole \; KClO_3}\right)$

$= 0.8446 \; g \; KClO_3$ (calculator answer)

$= 0.845 \; g \; KClO_3$ (correct answer)

17. $\dfrac{P_1 V_1}{T_1} = \dfrac{P_2 V_2}{T_2} \qquad V_2 = V_1 \times \dfrac{T_2}{T_1} \times \dfrac{P_1}{P_2}$

$P_1 = 748 \; torr - 28.3 \; torr$ (vp H_2O, at $28^\circ C$)

$= 719.7 \; torr$ (calculator answer)

$= 720 \; torr$ (correct answer)

$V_1 = 225 \; mL$

$T_1 = 28^\circ C + 273 = 301 \; K$

$P_2 = 760 \; torr$

$V_2 = ? \; mL$

$T_2 = 0^\circ C + 273 = 273 \; K$

$V_2 = 225 \; mL \times \left(\dfrac{273 \; K}{301 \; K}\right) \times \left(\dfrac{720 \; torr}{760 \; torr}\right) = 193.32925 \; mL$

(calculator answer)

$= 193 \; mL$ (correct answer)

ANSWERS TO MULTIPLE CHOICE EXERCISES

1. c
2. a
3. b
4. c
5. d
6. c
7. b
8. a
9. a
10. d
11. d

12. b
13. d
14. d
15. b
16. b
17. c
18. c
19. c
20. b
21. d

CHAPTER 13
Solutions: Terminology and Concentrations

REVIEW OF CHAPTER OBJECTIVES

1. <u>Define the terms solution, solute and solvent</u> (section 13.1).

    A solution is a homogeneous (uniform) mixture of two or more substances. In order to be homogeneous, the particles present must be of atomic and molecular size.

    A solution is composed of a solvent and one or more solutes. The solvent is the component of the solution which is present in the larger amount, and is the medium in which the other substances are dissolved. A solute is a component of a solution present in smaller amounts (than the solvent).

    Usually it is the solute which is the active ingredient in a solution; the solute usually undergoes reaction. Most often solutions are liquids, with the solvent usually being water.

2. <u>List the nine types of two-component solutions</u> (section 13.2).

    The nine types of two-component solutions are based on the physical states of the solvent and solute. They are:

    Gaseous solutions (where a gas is the solvent)

    > 1. gas in gas
    > 2. liquid in gas
    > 3. solid in gas

    Liquid solutions (where a liquid is the solvent)

    > 4. gas in liquid
    > 5. liquid in liquid
    > 6. solid in liquid

Solid solutions (where a solid is the solvent)

    7.   gas in solid
    8.   liquid in solid
    9.   solid in solid

The most common solutions are the ones in which the solvent is a liquid, and therefore the resulting solution (homogeneous mixture of solute in solvent) is a liquid. Most solutions encountered in a beginning chemistry course are of a solute dissolved in water, so, unless a solvent other than water is specified, the term solution will mean that the named solute is dissolved in water. As an example, a solution labeled "sodium chloride solution" contains sodium chloride (the solute) dissolved in water (the solvent).

3. <u>Define and distinguish between the terms saturated and unsaturated, dilute and concentrated, miscible and and immiscible</u> (section 13.2).

Solutions can be described as "saturated" or "unsaturated." A saturated solution contains the maximum amount of solute that can be dissolved under the conditions at which the solution exists. An unsaturated solution contains less than the maximum amount of solute that can be dissolved in the solution.

Solutions can be described as "dilute" or "concentrated." A dilute solution contains a small amount of solute relative to the amount that could dissolve. A concentrated solution contains a large amount of solute relative to the amount that could dissolve. A concentrated solution need not be a saturated solution (but it may be).

Mixtures of two liquids can be described as "miscible" or "immiscible." Miscible liquids, such as ethyl alcohol and water, are completely soluble in each other in all proportions. They form a single layer; a homogeneous mixture. Immiscible liquids are not soluble in each other. They form two layers, as carbon tetrachloride and water.

4. <u>Define the term solubility and know what is qualitatively meant by the terms insoluble, slightly soluble, soluble and very soluble</u> (section 13.2).

The solubility of a solute is the amount of a solute that will dissolve in a given amount of solvent. It is dependent on the nature of the solute and the solvent, and on the temperature. One way of expressing solubility is in grams of solute per 100 grams of solvent at a given temperature. Qualitatively, a solute is said to be:

insoluble, if less than 0.1 g of solute dissolves in 100 g solvent

slightly soluble, if between 0.1 g and 1 g of solute dissolves in 100 g solvent

soluble, if 1 g to 10 g of solute dissolves in 100 g solvent

very soluble, if more than 10 g of solute dissolves in 100 g solvent

As examples:

AgCl is insoluble in water -- only 0.000089 g AgCl dissolves in 100 g water at 10°C

$PbBr_2$ is slightly soluble in cold water -- 0.84 g $PbBr_2$ dissolves in 100 g H_2O at 20°C

$PbCl_2$ is soluble in hot water -- 3.34 g $PbCl_2$ dissolves in 100 g H_2O at 100°C

$Pb(NO_3)_2$ is very soluble in cold water -- 56.5 g $Pb(NO_3)_2$ dissolves in 100 g H_2O at 20°C

5. <u>Describe the solution process at a molecular level as an ionic solute dissolves in water</u> (section 13.3).

In a solution, solute particles are uniformly dispersed throughout the solvent. For a solute to dissolve in a solvent two types of interparticle attractions must be overcome:

1. attractions between solute particles (solute–solute attractions), and

2. attractions between solvent particles (solvent–solvent attractions).

A solute dissolves (in a solvent) if the attraction between solute and solvent particles (solute–solvent attractions) are greater than solute–solute and solvent–solvent attractions.

When an ionic solute dissolves in water, the polar water molecules surround the ions of the solute, with the negative oxygen end of the water molecules attracting the positive ions, and the positive hydrogen end of the water molecules attracting the negative ions. This causes the ions to break away from the surface of the crystal. Once "free" they are surrounded by water molecules (hydrated). As new ions (on the surface of the crystal) are exposed to water, they too become hydrated and go into solution.

6. <u>Apply the solubility rule "likes dissolve likes" and use the solubility guidelines for ionic solutes in water</u> (section 13.4).

A very useful generalization for predicting solubilities is "substances of like polarity tend to be more soluble in each other than substances which differ in polarity." More simply stated we can say "likes dissolve likes." This general rule of "likes dissolve likes" works well for all except ionic solutes. Not all ionic solutes are soluble in the polar liquid water as the rule would predict. The charge and the size of ions are complicating factors in predicting ionic solute solubilities. For ionic solutes the table below gives better solubility guidelines than "likes dissolve likes."

Ion in the compound	Solubility in Water	Exceptions
Group IA (Li^+, Na^+, etc.) and ammonium $(NH_4)^+$	soluble	
Acetate ($C_2H_3O_2^-$) and nitrate ($NO_3)^-$	soluble	
Cl^-, Br^-, I^-	soluble	Ag^+, Pb^{2+}, Hg_2^{2+}
SO_4^{2-}	soluble	Ca^{2+}, Sr^{2+}, Ba^{2+}, Pb^{2+}
CO_3^{2-}, PO_4^{3-}	insoluble	Group IA & NH_4^+
S^{2-}	insoluble	Group IA and IIA, NH_4^+
OH^-	insoluble	Group IA, Ba^{2+}, Sr^{2+}

Example: Predict the solubility of the following solutes in water:

a) CCl_4 (nonpolar), b) AgI, c) $(NH_4)_2CO_3$,
d) Na_2S

Solution:

a) Insoluble. A nonpolar substance is not soluble in a polar solvent (water).

b) Insoluble. All I^- salts are soluble except those of Ag^+, Pb^{2+} and Hg_2^{2+}.

c) Soluble. All NH_4^+ salts are soluble.

d) Soluble. All Na^+ salts are soluble.

It is important to learn these guidelines for the solubilities of ionic compounds in water. They will become very important in the next chapter when net ionic equations are discussed.

7. <u>Define the term concentration, and distinguish between the terms concentration and solubility</u> (section 13.5).

The concentration of a solution is the amount of solute present in a given amount of solvent or solution.

$$\text{Concentration of a solution} = \frac{\text{amount of solute}}{\text{amount of solvent}}$$

or

$$\text{Concentration of a solution} = \frac{\text{amount of solute}}{\text{amount of solution}}$$

The solubility of a solute is how much solute <u>could</u> dissolve in a given amount of solvent, whereas the concentration of a solution is how much solute <u>is</u> dissolved in a given amount of solvent (or solution).

Several different sets of units are used to express solution concentration.

8. <u>Give the defining equation for the solution concentrations percent by weight, percent by volume and weight-volume percent</u> (section 13.6).

Percent concentration -- Percent means parts per hundred.

Since amounts of solute and solution can be stated in either mass or volume, we can describe solution concentration in:

1) weight-weight percent (or percent by weight)
2) volume-volume percent (or percent by volume)
3) weight-volume percent

Weight-weight percent (or percent by weight) is the percentage unit most often used by chemists and is the one indicated if neither weight nor volume is stated. If a solution is labeled "15% NaOH solution" it is a solution which contains 15% by mass of NaOH, or 15 g NaOH in every 100 g solution.

$$\text{Percent by weight} = \frac{\text{mass solute}}{\text{mass solution}} \times 100$$

The 15% NaOH solution, which contains 15 g NaOH in 100 g solution, is prepared by mixing 15 g NaOH + 85 g H_2O (to make 100 g solution).

Example: Calculate the percent by weight concentration of $KMnO_4$ in a solution made by mixing 8.5 g $KMnO_4$ with enough water to make 62 g solution.

Solution:

$$\% \text{ by weight} = \frac{\text{mass solute}}{\text{mass solution}} \times 100$$

$$= \frac{8.5 \text{ g } KMnO_4}{62 \text{ g solution}} \times 100$$

$$= 13.709677\% \ KMnO_4 \text{ (calculator answer)}$$

$$= 14\% \ KMnO_4 \text{ (correct answer)}$$

Example: How many grams of $K_2Cr_2O_7$ are in 500 g of 25.0% by weight $K_2Cr_2O_7$ solution?

Solution:

Using dimensional analysis and the conversion factor 25.0 g $K_2Cr_2O_7$ = 100 g solution (determined from the percent concentration given), we get:

$$500 \text{ g solution} \times \left(\frac{25.0 \text{ g } K_2Cr_2O_7}{100 \text{ g solution}} \right)$$

$$= 125 \text{ g } K_2Cr_2O_7 \text{ (calculator and correct answer)}$$

Volume-volume percent (or percent by volume) is used when both solute and solvent are liquids or gases, as volume is often easier to measure than mass.

$$\text{Percent by volume} = \frac{\text{volume solute}}{\text{volume solution}} \times 100$$

A 15% by volume solution of ethyl alcohol (in water) contains 15 mL solute (ethyl alcohol) in 100 mL solution.

Weights are additive:

12 g solute + 88 g solvent = 100 g solution

Volumes are not additive:

50 mL water + 50 mL ethyl alcohol = 97 mL solution

Thus, the total volume of solution must be known when doing percentage by volume calculations.

Example: What is the percent by volume of a solution containing 25 mL ethyl alcohol in a total volume of 45 mL?

Solution:

$$\text{Percent by volume} = \frac{\text{volume solute}}{\text{volume solution}} \times 100$$

$$= \frac{25 \text{ mL ethyl alcohol}}{45 \text{ mL solution}} \times 100$$

= 55.555556% (calculator answer)
= 56% (correct answer)

Example: How many mL of methyl alcohol are needed to make 5.5 mL of 65% by volume methyl alcohol solution?

Solution:

$$5.5 \text{ mL solution} \times \left(\frac{65 \text{ mL methyl alcohol}}{100 \text{ mL solution}} \right)$$

= 3.575 mL (calculator answer)
= 3.6 mL (correct answer)

Weight-volume percent is often useful when the solute is a solid (easily weighed) and the solvent is a liquid (whose volume is easily measured).

$$\text{weight-volume percent} = \frac{\text{g solute}}{\text{mL solution}} \times 100$$

Example: How many grams of KNO_3 are needed to make $\overline{5000}$ mL of a 10.0% weight-volume percent KNO_3 solution?

Solution:

$$\overline{5000}~\cancel{\text{mL solution}} \times \left(\frac{10.0~\text{g } KNO_3}{100~\cancel{\text{mL solution}}} \right)$$

$$= 500~\text{g } KNO_3~\text{(calculator answer)}$$

$$= \overline{500}~\text{g } KNO_3~\text{(correct answer)}$$

9. <u>Give the defining equation for molarity and use it to calculate any one of the quantities in the equation (given the other two)</u> (section 13.7).

Laboratory solutions are often labeled with the concentration term molarity. The molarity of a solution (M) tells the number of moles of solute per liter of solution.

$$M = \frac{\text{moles solute}}{\text{liters solution}}$$

Example: Calculate the molarity of a solution containing 1.3 grams of $KMnO_4$ in 25 mL solution.

Solution:

The volume of solution is given, but not in the right unit. Converting to liters, we have

$$25~\cancel{\text{mL}} \times \left(\frac{1~\text{L}}{1000~\cancel{\text{mL}}} \right) = 0.025~\text{L (calculator and correct answer)}$$

The moles of solute present is calculated from the grams of solute:

$$1.3~\cancel{\text{g } KMnO_4} \times \left(\frac{1~\text{mole } KMnO_4}{158~\cancel{\text{g } KMnO_4}} \right)$$

$$= 0.00822785~\text{moles } KMnO_4~\text{(calculator answer)}$$

$$= 0.0082~\text{moles } KMnO_4~\text{(correct answer)}$$

The molarity of the solution is:

$$M = \frac{0.0082 \text{ moles KMnO}_4}{0.025 \text{ L solution}}$$

$$= 0.328 \text{ M (calculator answer)}$$
$$= 0.33 \text{ M (correct answer)}$$

Example: How many liters of 6.00 M Na_2SO_4 solution can be prepared from 500 g Na_2SO_4?

Solution:

The molarity of the solution is used as a conversion factor in the dimensional analysis set-up of this problem.

$$500 \text{ g Na}_2\text{SO}_4 \text{ x} \left(\frac{1 \text{ mole Na}_2\text{SO}_4}{142 \text{ g Na}_2\text{SO}_4} \right) \text{x} \left(\frac{1 \text{ L Na}_2\text{SO}_4}{6.00 \text{ mole Na}_2\text{SO}_4} \right)$$

$$= 0.58685446 \text{ L Na}_2\text{SO}_4 \text{ (calculator answer)}$$
$$= 0.587 \text{ L Na}_2\text{SO}_4 \text{ (correct answer)}$$

10. Give the defining equation for molaltiy, and use it to calculate any one of the quantities in the equation when the other two are given (section 13.8).

The molality of a solution, m, gives the number of moles of solute per kilogram of solvent. Note that it is kilograms of solvent rather than kilograms of solution.

$$(\text{molality}) \text{ m} = \frac{\text{moles solute}}{\text{kg solvent}}$$

Molality (m) is the preferred concentration term where changes in temperature are involved because kg solvent is independent of temperature (while liters of solution varies with temperature).

Example: Calculate the molality of a solution containing 10.0 grams of NaOH in 500 g of water.

Solution:

The mass of solution is given, but in the wrong units. Converting to kilograms, we have

$$500 \text{ g H}_2\text{O} \text{ x} \left(\frac{1 \text{ kg H}_2\text{O}}{1000 \text{ g H}_2\text{O}} \right) = 0.5 \text{ kg H}_2\text{O}$$
$$\text{(calculator answer)}$$
$$= 0.500 \text{ kg H}_2\text{O}$$
$$\text{(correct answer)}$$

The moles of solute present is calculated from the grams of solute

$$10.0 \text{ g NaOH} \times \left(\frac{1 \text{ mole NaOH}}{40.0 \text{ g NaOH}}\right) = 0.25 \text{ moles NaOH}$$
(calculator answer)
$$= 0.250 \text{ moles NaOH}$$
(correct answer)

The molality of the solution is

$$m = \frac{0.250 \text{ moles NaOH}}{0.500 \text{ kg H}_2\text{O}} = 0.5 \text{ m (calculator answer)}$$
$$= 0.500 \text{ m (correct answer)}$$

Example: Calculate the number of grams of $KMnO_4$ which must be added to 200 g water to prepare a 0.100 m solution.

Solution:

$$200 \text{ g water} \times \left(\frac{1 \text{ kg water}}{1000 \text{ g water}}\right) \times \left(\frac{0.100 \text{ mole KMnO}_4}{1 \text{ kg water}}\right)$$

$$\times \left(\frac{158 \text{ g KMnO}_4}{1 \text{ mole KMnO}_4}\right) = 3.16 \text{ g KMnO}_4 \text{ (calculator and correct answer)}$$

Note how the concentration of the solution (0.100 m) was used as a conversion factor (the second one).

11. <u>Give the defining equation for normality and find any one of the quantities when the other two are given</u> (section 13.9).

An equivalent of an ionic solute is the quantity of the solute which produces one mole of positive charge (or one mole of negative charge) upon complete dissociation of the solute. The equivalent weight of a substance is the mass, in grams, of one equivalent of the substance.

Example: Calculate the equivalent weight of:

a) $KMnO_4$ (molecular weight = 158 amu)

b) K_2SO_4 (molecular weight = 174 amu)

c) Na_3PO_4 (molecular weight = 164 amu)

Solution:

a) $KMnO_4 = K^+ + MnO_4^-$

One mole of $KMnO_4$ dissociates to give one mole of K^+, and one mole of MnO_4^-. Therefore, 1 equivalent = 1 mole, and the equivalent weight of $KMnO_4$ is 158.

b) $K_2SO_4 = 2K^+ + SO_4^{2-}$

One mole of K_2SO_4 dissociates to produce 2 moles of positive (and negative) charge. Therefore, one mole K_2SO_4 = 2 equivalents of K_2SO_4.

$$\left(\frac{174 \text{ g } K_2SO_4}{1 \text{ mole } K_2SO_4}\right) \times \left(\frac{1 \text{ mole } K_2SO_4}{2 \text{ equivalents } K_2SO_4}\right)$$

$$= 87 \text{ g } K_2SO_4/\text{equivalent (calculator answer)}$$
$$= 87.0 \text{ g } K_2SO_4/\text{equivalent (correct answer)}$$

c) $Na_3PO_4 = 3Na^+ + PO_4^{3-}$

One mole of Na_3PO_4 dissociates to produce 3 moles of positive (and negative) charge. Therefore, one mole Na_3PO_4 = 3 equivalents of Na_3PO_4 .

$$\left(\frac{164 \text{ g } Na_3PO_4}{1 \text{ mole } Na_3PO_4}\right) \times \left(\frac{1 \text{ mole } Na_3PO_4}{3 \text{ equivalents } Na_3PO_4}\right)$$

$$= 51.333333 \text{ g } Na_3PO_4/\text{equivalent (calculator answer)}$$
$$= 51.3 \text{ g } Na_3PO_4/\text{equivalent (correct answer)}$$

Example: Calculate the normality of a solution made by dissolving 25.0 g H_2SO_4 in 0.250 L water.

Solution:

The dissociation equation is

$$H_2SO_4 \longrightarrow 2H^+ + SO_4^{2-}$$

Thus, 2 moles of positive charge and 2 moles of negative charge results from the dissociation of 1 mole of compound. This means that

$$1 \text{ mole } H_2SO_4 = 2 \text{ equivalents } H_2SO_4$$

The number of equivalents of H_2SO_4 is:

$$25.0 \text{ g } H_2SO_4 \times \left(\frac{1 \text{ mole } H_2SO_4}{98.1 \text{ g } H_2SO_4}\right) \times \left(\frac{2 \text{ equiv } H_2SO_4}{1 \text{ mole } H_2SO_4}\right)$$

$$= 0.509684 \text{ equivalent } H_2SO_4 \text{ (calculator answer)}$$

$$= 0.510 \text{ equivalent } H_2SO_4 \text{ (correct answer)}$$

The normality of the solution is:

$$N = \frac{0.510 \text{ equiv } H_2SO_4}{0.250 \text{ L solution}} = 2.04 \text{ N (calculator and correct answer)}$$

12. <u>Calculate the molarity of a solution prepared by diluting a stock solution of known molarity</u> (section 13.10).

Dilution is the process of adding solvent to a solution to decrease its concentration. On adding water to a solution, the same number of moles of solute remain, but they are in a larger volume of solution so the concentration is lower. The number of moles of solute remains unchanged when diluting. The equation:

$$M_1V_1 = M_2V_2$$

is very useful in problems where dilution occurs. Its derivation is given on page 364 of the textbook.

Example: Calculate the molarity of a solution prepared by diluting 42.0 mL of 6.00 M HCl solution to 500 mL.

Solution:

Three of the four quantities in the equation

$$M_1V_1 = M_2V_2$$

are known. Hence the fourth can be calculated.

$M_1 = 6.00 \text{ M}$ $\qquad M_2 = ?$

$V_1 = 42.0 \text{ mL}$ $\qquad V_2 = 500 \text{ mL}$

$$M_2 = \frac{M_1V_1}{V_2} = \frac{(6.00 \text{ M}) \times (42.0 \text{ mL})}{(500 \text{ mL})}$$

$$= 5.04 \text{ M (calculator and correct answer)}$$

Example:　How many mL of 6.00 M Na_2SO_4 solution are
needed to prepare $15\overline{0}$ mL of 0.300 M Na_2SO_4
solution?

Solution:

Three of the four quantities in the equation

$$M_1V_1 = M_2V_2$$

are known.

M_1 = 6.00 M　　　　　　　M_2 = 0.300 M

V_1 = ?　　　　　　　　　V_2 = $15\overline{0}$ mL

Solving for V_1 gives

V_1 = $\dfrac{(15\overline{0} \text{ mL}) \text{ x } (0.300 \text{ M})}{(6.00 \text{ M})}$ = 7.5 mL
　　　　　　　　　　　　　　　　　(calculator answer)
　　　　　　　　　　　　= 7.50 mL (correct answer)

13. Calculate [given a balanced equation, the volume and
    molarity of one reactant and the volume (or molarity) of
    a second reactant] the molarity (or volume) of the second
    reactant (section 13.11).

    When solution concentrations are expressed in terms of
    molarity a direct relationship exists between the volume
    of the solution (in liters) and the number of moles of
    solute present. The definition of molarity gives this
    relationship. Since

    $$M = \frac{\text{moles solute}}{\text{liters solution}}$$

    this relationship can be incorporated into problem
    solving situations as diagrammed on page 367 in your
    textbook.

    Example:　How many liters of 0.555 M NaOH are needed to
    react with 0.0285 L of 0.855 M H_3PO_4
    according to the equation:

    $$3NaOH + H_3PO_4 \longrightarrow Na_3PO_4 + 3H_2O$$

    Solution:

    In terms of Figure 13-4 on page 367 of the textbook
    this is a volume-to-volume problem. The pathway is:

$$\begin{array}{ccccc} \text{Volume} & & \text{Moles} & & \text{Moles} & & \text{Volume} \\ \text{H}_3\text{PO}_4 & \rightarrow & \text{H}_3\text{PO}_4 & \rightarrow & \text{NaOH} & \rightarrow & \text{NaOH} \end{array}$$

Using dimensional analysis the set-up is:

$$0.0285 \ \text{L H}_3\text{PO}_4 \ \times \left(\frac{0.855 \ \text{moles H}_3\text{PO}_4}{1 \ \text{L H}_3\text{PO}_4} \right) \times \left(\frac{3 \ \text{moles NaOH}}{1 \ \text{mole H}_3\text{PO}_4} \right)$$

$$\times \left(\frac{1 \ \text{L NaOH}}{0.555 \ \text{mole NaOH}} \right) \begin{array}{l} = 0.13171622 \ \text{L NaOH} \\ \text{(calculator answer)} \\ = 0.132 \ \text{L NaOH} \\ \text{(correct answer)} \end{array}$$

14. <u>Calculate, given a balanced chemical equation and the volume and molarity of one reactant combining with an excess of other reactants, the mass or volume (for gases) of any product that is produced</u> (section 13.11).

Figure 13-4 on page 367 in the textbook also serves as a "roadmap" for solving problems of the type mentioned in this objective.

Example: How many liters of Cl_2 gas, at STP, can be produced from 0.0357 L of 0.854 M HCl solution according to the reaction:

$$\text{K}_2\text{Cr}_2\text{O}_7 + 14\text{HCl} \longrightarrow 2\text{KCl} + 2\text{CrCl}_3 + 7\text{H}_2\text{O} + 3\text{Cl}_2$$

Solution:

In terms of Figure 13-4 this is a volume (solution)-to-volume (gas) problem. The pathway, using Figure 13-4, is:

$$\begin{array}{ccccc} \text{Volume} & & \text{Moles} & & \text{Moles} & & \text{Volume} \\ \text{HCl} & \rightarrow & \text{HCl} & \rightarrow & \text{Cl}_2 & \rightarrow & \text{Cl}_2 \end{array}$$

The set-up is:

$$0.0357 \ \text{liters HCl solution} \ \times \left(\frac{0.854 \ \text{moles HCl}}{1 \ \text{liter HCl solution}} \right)$$

$$\times \left(\frac{3 \ \text{moles Cl}_2}{14 \ \text{moles HCl}} \right) \times \left(\frac{22.4 \ \text{liters Cl}_2}{1 \ \text{mole Cl}_2} \right) \begin{array}{l} = 0.14634144 \ \text{L Cl}_2 \\ \text{(calculator answer)} \\ = 0.146 \ \text{L Cl}_2 \\ \text{(correct answer)} \end{array}$$

PROBLEM SET

1. A bottle is labeled 15.0% by weight $K_2Cr_2O_7$ solution. The solute is _____, the solvent is _____ and 250 g of this 15.0% $K_2Cr_2O_7$ solution contains how many grams of solute?

2. Describe the appearance of each of the following mixtures after the 2 substances have been shaken together:

   a) 3.0 g of $Ca(HCO_3)_2$ and 200 g H_2O. $Ca(HCO_3)_2$ is very soluble in water.

   b) 5.0 g of $CaCO_3$ and 200 grams of water. $CaCO_3$ is insoluble in water.

   c) 50 mL of ethyl alcohol and 50 mL of water. Ethyl alcohol and water are miscible.

   d) 50 mL CCl_4 and 50 mL water. CCl_4 and water are immiscible.

3. Calculate the concentration of the following solutions as percent by weight:

   a) 2.85 g $KMnO_2$ in 85.0 g solution

   b) 1.50 g NaCl plus 25.0 g water

   c) 2.85 moles of NaOH in 2000 g solution

4. Calculate the molarity of the following solutions:

   a) 0.0250 moles $K_2Cr_2O_7$ in 100 mL solution

   b) 288 g $KMnO_4$ in 5.00 liters solution

   c) 17.5 g $CaCl_2$ in 500 mL solution

5. Calculate the number of grams of solute in:

   a) 2.00 kg of 7.00% by weight NaCl solution

   b) 586 grams of 17.0% by weight $K_2Cr_2O_7$ solution

   c) 3.50 liters of 1.50 M $Mg(NO_3)_2$ solution

6. Calculate the molality of a solution containing:

    a) 2.75 moles $KMnO_4$ in $\overline{500}$ g water

    b) 287 g KOH in 2.00 kg water

7. A solution is made by mixing 25.0 g $CaCl_2$ with $\overline{500}$ g water.

    a) Calculate the % by weight concentration of this solution

    b) Calculate the molality of this solution

8. Calculate the normality of a solution:

    a) containing 2.50 moles of K_2SO_4 in $\overline{500}$ mL solution

    b) containing 18.7 grams of $NaNO_3$ in 2.50 liters solution

    c) labeled "2.78 M $(NH_4)_3PO_4$ solution"

9.   a) How many <u>mL</u> of 5.00 M HCl solution is needed to prepare $\overline{300}$ mL of 0.650 M HCl solution?

    b) What volume of 0.520 M HCl solution can be prepared from 27.3 mL of 1.00 M HCl solution?

    c) What is the molarity of a solution made by mixing 125.0 mL of water with 50.0 mL of 2.75 M KOH solution?

10. Solid $NaHCO_3$ and H_2SO_4 solution react according to the following equation:

$$2NaHCO_{3(s)} + H_2SO_{4(aq)} \longrightarrow Na_2SO_{4(aq)} + 2CO_{2(g)} + 2H_2O_{(\ell)}$$

    a) How many mL of 0.520 M H_2SO_4 solution is needed to react with 2.75 g $NaHCO_3$?

    b) 25.0 mL of 0.555 M H_2SO_4 solution produces how many mL of CO_2 gas at STP?

    c) What is the molarity of a H_2SO_4 solution if 32.7 mL of it is required to react with 1.85 g of $NaHCO_3$?

11. How many grams of Ag_3PO_4 are produced by the reaction of 25.0 mL of 0.254 M $AgNO_3$ solution and excess H_3PO_4 according to the equation:

$$3AgNO_{3(aq)} + H_3PO_{4(aq)} \longrightarrow Ag_3PO_{4(s)} + 3HNO_{3(aq)}$$

MULTIPLE CHOICE EXERCISES

1. A solution may contain:

   a) only one solvent but many solutes
   b) many solvents but only one solute
   c) only one solvent and only one solute
   d) many solvents and many solutes

2. In a saturated solution,

   a) undissolved solute must be present
   b) no undissolved solute may be present
   c) the solubility limit for the solute has been reached
   d) solid crystallizes out if the solution is stirred

3. The term slightly soluble indicates that the solubility of a solute per 100 grams of solvent is:

   a) less than 0.1 gram
   b) between 0.1 and 1 gram
   c) between 1 and 10 grams
   d) between 10 and 20 grams

4. Which of the following statements concerning dilute solutions is correct?

   a) all dilute solutions are unsaturated
   b) dilute solutions may be saturated but need not be
   c) a dilute solution must contain less than 1 g solute
   d) a dilute solution can never contain more than 20 g solute per 100 g solvent

5. When NaCl dissolves in water, the sodium ions (Na^+) are:

   a) attracted to the oxygen ends of H_2O molecules
   b) attracted to the hydrogen ends of H_2O molecules
   c) bonded covalently to the H_2O molecules
   d) completely ignored by the H_2O molecules

6. Consider the following substances and their polarities: A -- polar, B -- polar, C -- nonpolar, D -- nonpolar. It is true that:

   a) A will be more soluble in C than in B
   b) C will be more soluble in D than in A
   c) B will be more soluble in D than in A
   d) D will be more soluble in B than in C

7. Which of the following ionic compounds is not soluble in water?

a) $Mg(NO_3)_2$
b) $NaCl$
c) $(NH_4)_2SO_4$
d) $CaCO_3$

8. Which of the following ionic compounds is not soluble in water?

a) $NaC_2H_3O_2$
b) Ag_3PO_4
c) K_2CO_3
d) $(NH_4)_3PO_4$

9. The molarity of a solution is defined as the number of:

a) grams of solute per liter of solution
b) moles of solute per liter of solution
c) moles of solute per liter of solvent
d) moles of solute per kilogram of solvent

10. Which of the following concentration units has a defining equation that contains the quantity "kilograms of solvent?"

a) molality
b) normality
c) percentage (w/v)
d) percentage (v/v)

11. Which of the following concentration units has a defining equation that contains the quantity "equivalents of solute?"

a) molality
b) normality
c) percentage (w/v)
d) percentage (v/v)

12. A solution containing 8 grams of NaCl in 400 grams of solution has a percentage concentration (w/w) of:

a) 0.5%   c) 4%
b) 2%   d) 20%

Solutions: Terminology and Concentrations   277

13. The number of grams of NaOH (mol. wt. = 40) required to prepare 2.00 liters of 1.00 M solution is:

    a) 10 grams
    b) 20 grams
    c) 40 grams
    d) 80 grams

14. Molality could be used as a conversion factor between:

    a) grams of solute and moles of solvent
    b) moles of solute and volume of solution
    c) grams of solute and volume of solution
    d) moles of solute and kilograms of solvent

15. Which of the following would serve as a direct conversion factor from grams of solution to grams of solute?

    a) percent by weight
    b) molality
    c) molarity
    d) percent by volume

16. The number of equivalents of NaOH in 2.0 liters of 5 N NaOH solution is:

    a) 2 equivalents
    b) 4 equivalents
    c) 10 equivalents
    d) 12 equivalents

17. For the compound MgF_2 the number of equivalents in a mole will be:

    a) 1
    b) 2
    c) 3
    d) 4

18. What is the normality of a solution made by dissolving 213.0 grams of $Al(NO_3)_3$ (mol. wt. = 213.0) in enough water to give 0.50 liters of solution?

    a) 1.0 N
    b) 2.13 N
    c) 3.0 N
    d) 6.0 N

19. What is the concentration of a solution made by diluting 20 mL of 3.0 M NaOH with 40 mL of H_2O?

    a) 1.0 M
    b) 1.8 M
    c) 2.0 M
    d) 6.0 M

20. If 200 mL of 2.00 M NaOH is diluted to 500 mL, the resulting solution contains:

    a) 0.400 moles NaOH
    b) 2.00 moles NaOH
    c) 4.00 moles NaOH
    d) 40.0 moles NaOH

21. How many liters of 2.00 M HCl are needed to completely react with 1.00 mole of $CaCO_3$ according to the equation:

$$CaCO_3 + 2HCl \longrightarrow CaCl_2 + H_2O + CO_2$$

    a) 0.500 L
    b) 1.00 L
    c) 1.50 L
    d) 2.00 L

ANSWERS TO PROBLEMS SET

1. 15.0% by weight $K_2Cr_2O_7$ solution contains 15.0 g $K_2Cr_2O_7$ (the solute) and 85.0 g H_2O (the solvent) in every 100.0 g of solution.

    250 g solution x $\left(\dfrac{15.0 \text{ g solute}}{100 \text{ g solution}}\right)$ = 37.5 g solute (calculator and correct answer)

2. a) Since $Ca(HCO_3)_2$ is very soluble in water (over 10 g of this solute would dissolve in 100 g water), this would be a homogeneous mixture. There would be only one layer, a solution.

    b) Since $CaCO_3$ is insoluble, less than 1 g of solute would dissolve in 100 g water, and 2 layers would be seen in the mixture -- undissolved solid $CaCO_3$ and the water layer.

c) Ethyl alcohol and water are miscible; they form one
layer, a solution (homogeneous mixture).

d) CCl_4 and water are immiscible; after shaking, two
separate layers are seen in the mixture.

3. a) $\dfrac{2.85 \text{ g solute}}{85 \text{ g solution}} \times 100 = 3.3529412\%$ (calculator answer)
$= 3.35\%$ (correct answer)

b) $\dfrac{1.50 \text{ g solute}}{(1.50 + 25.0) \text{ g solution}} \times 100 = 5.6603774\%$
(calculator answer)
$= 5.66\%$ (correct answer)

c) $\left(\dfrac{2.85 \text{ moles NaOH}}{2000 \text{ g solution}}\right) \times \left(\dfrac{40.0 \text{ g NaOH}}{1 \text{ mole NaOH}}\right) \times 100$

$= 5.7\%$ (calculator answer)
$= 5.70\%$ (correct answer)

4. a) $M = \dfrac{\text{moles solute}}{\text{liters solution}} = \left(\dfrac{0.0250 \text{ moles}}{100 \text{ mL}}\right) \times \left(\dfrac{1000 \text{ mL}}{1 \text{ liter}}\right)$

$= 0.25$ M (calculator answer)

$= 0.250$ M (correct answer)

b) $M = \left(\dfrac{288 \text{ g KMnO}_4}{5.00 \text{ liters}}\right) \times \left(\dfrac{1 \text{ mole KMnO}_4}{158 \text{ g KMnO}_4}\right) = 0.36455696$ M
(calculator answer)
$= 0.365$ M
(correct answer)

c) $\left(\dfrac{17.5 \text{ g CaCl}_2}{500 \text{ mL}}\right) \times \left(\dfrac{1000 \text{ mL}}{1 \text{ liter}}\right) \times \left(\dfrac{1 \text{ mole CaCl}_2}{111 \text{ g CaCl}_2}\right)$

$= 0.31531532$ M (calculator answer)
$= 0.315$ M (correct answer)

5. a) $2.00 \text{ kg solution} \times \left(\dfrac{1000 \text{ g solution}}{1 \text{ kg solution}}\right) \times \left(\dfrac{7.00 \text{ g solute}}{100 \text{ g solution}}\right)$

$= 140$ g solute (calculator answer)
$= 14\overline{0}$ g solute (correct answer)

b) $586 \text{ g solution} \times \left(\dfrac{17.0 \text{ g solute}}{100 \text{ g solution}}\right) = 99.62$ g solute
(calculator answer)
$= 99.6$ g solute
(correct answer)

c) $3.50 \text{ liters} \times \left(\dfrac{1.50 \text{ mole Mg(NO}_3)_2}{1 \text{ liter}}\right) \times \left(\dfrac{148 \text{ g Mg(NO}_3)_2}{1 \text{ mole Mg(NO}_3)_3}\right)$

$= 777$ g $Mg(NO_3)_2$ (calculator and correct answer)

6. a) $m = \dfrac{\text{moles solute}}{\text{kg solvent}} = \left(\dfrac{2.75 \text{ moles KMnO}_4}{500 \text{ g } \cancel{\text{H}_2\text{O}}}\right) \times \left(\dfrac{1000 \text{ g } \cancel{\text{H}_2\text{O}}}{1 \text{ kg H}_2\text{O}}\right)$

$= 5.5 \text{ m (calculator answer)}$
$= 5.50 \text{ m (correct answer)}$

b) $m = \dfrac{\text{moles solute}}{\text{kg solvent}} = \left(\dfrac{287 \text{ g } \cancel{\text{KOH}}}{2.00 \text{ kg H}_2\text{O}}\right) \times \left(\dfrac{1 \text{ mole KOH}}{56.1 \text{ g } \cancel{\text{KOH}}}\right)$

$= 2.5579323 \text{ m (calculator answer)}$
$= 2.56 \text{ m (correct answer)}$

7. a) $\% = \dfrac{\text{g solute}}{\text{g solution}} \times 100 = \dfrac{25.0 \text{ g solute}}{(25.0 + 500) \text{ g solution}} \times 100$

$= 4.7619048\% \text{ (calculator answer)}$
$= 4.76\% \text{ (correct answer)}$

b) $m = \dfrac{\text{moles solute}}{\text{kg solvent}} = \left(\dfrac{25.0 \text{ g } \cancel{\text{CaCl}_2}}{500 \text{ g } \cancel{\text{H}_2\text{O}}}\right) \times \left(\dfrac{1 \text{ mole CaCl}_2}{111 \text{ g } \cancel{\text{CaCl}_2}}\right)$

$\times \left(\dfrac{1000 \text{ g } \cancel{\text{H}_2\text{O}}}{1 \text{ kg H}_2\text{O}}\right) = 0.45045045 \text{ m (calculator answer)}$
$= 0.450 \text{ m (correct answer)}$

8. a) $N = \dfrac{\text{equivalents}}{\text{liter}} = \left(\dfrac{2.50 \text{ moles } \cancel{\text{K}_2\text{SO}_4}}{500 \text{ mL}}\right) \times \left(\dfrac{1000 \text{ mL}}{1 \text{ liter}}\right)$

$\times \left(\dfrac{2 \text{ equivalents}}{1 \text{ mole } \cancel{\text{K}_2\text{SO}_4}}\right) = 10 \text{ N (calculator answer)}$
$= 10.0 \text{ N (correct answer)}$

b) $N = \dfrac{\text{equivalents}}{\text{liter}} = \left(\dfrac{18.7 \text{ g } \cancel{\text{NaNO}_3}}{2.50 \text{ liters}}\right) \times \left(\dfrac{1 \text{ mole } \cancel{\text{NaNO}_3}}{85.0 \text{ g } \cancel{\text{NaNO}_3}}\right)$

$\times \left(\dfrac{1 \text{ equivalent}}{1 \text{ mole } \cancel{\text{NaNO}_3}}\right) = 0.088 \text{ N (calculator answer)}$
$= 0.0880 \text{ N (correct answer)}$

c) $N = \dfrac{\text{equivalents}}{\text{liter}} = \left(\dfrac{2.78 \text{ mole } \cancel{(\text{NH}_4)_3\text{PO}_4}}{1 \text{ liter}}\right)$

$\times \left(\dfrac{3 \text{ equivalents}}{1 \text{ mole } \cancel{(\text{NH}_4)_3\text{PO}_4}}\right) = 8.34 \text{ N (calculator and}$
correct answer)

9. a) $M_1V_1 = M_2V_2 \qquad V_1 = \dfrac{V_2M_2}{M_1} = \dfrac{300 \text{ mL} \times 0.650 \cancel{M}}{5.00 \cancel{M}}$

$= 39 \text{ mL (calculator answer)}$
$= 39.0 \text{ mL (correct answer)}$

b) $M_1V_1 = M_2V_2$ $V_2 = \dfrac{V_1M_1}{M_2} = \dfrac{27.3 \text{ mL} \times 1.00 \text{ M}}{0.520 \text{ M}}$

$\qquad\qquad\qquad\qquad\qquad = 52.5$ mL (calculator and correct answer)

c) 125.0 mL H_2O + 50.0 mL solution = 175.0 mL solution

$M_1V_1 = M_2V_2$ $M_2 = \dfrac{M_1V_1}{V_2} = \dfrac{2.75 \text{ M} \times 50.0 \text{ mL}}{175 \text{ mL}}$

$\qquad\qquad\qquad\qquad\qquad = 0.78571429$ M (calculator answer)
$\qquad\qquad\qquad\qquad\qquad = 0.786$ M (correct answer)

10. a) g $NaHCO_3 \longrightarrow$ moles $NaHCO_3 \longrightarrow$ moles $H_2SO_4 \longrightarrow$

$\qquad\qquad\qquad\qquad\qquad\qquad\qquad$ liters $H_2SO_4 \longrightarrow$ mL H_2SO_4

$\qquad$ 75 g $NaHCO_3 \times \left(\dfrac{1 \text{ mole } NaHCO_3}{84.0 \text{ g } NaHCO_3}\right) \times \left(\dfrac{1 \text{ mole } H_2SO_4}{2 \text{ moles } NaHCO_3}\right)$

$\qquad\qquad\qquad\qquad \times \left(\dfrac{1 \text{ L } H_2SO_4}{0.520 \text{ mole } H_2SO_4}\right) \times \left(\dfrac{1000 \text{ mL } H_2SO_4}{1 \text{ L } H_2SO_4}\right)$

$\qquad\quad = 31.478938$ mL H_2SO_4 solution (calculator answer)
$\qquad\quad = 31.5$ mL H_2SO_4 solution (correct answer)

b) Volume H_2SO_4 solution $\longrightarrow$ moles $H_2SO_4 \longrightarrow$ moles CO_2

$\qquad\qquad\qquad\qquad\qquad\qquad\qquad \longrightarrow$ liters $CO_2 \longrightarrow$ mL CO_2

$\qquad$ 25.0 mL $H_2SO_4 \times \left(\dfrac{1 \text{ liter } H_2SO_4}{1000 \text{ mL } H_2SO_4}\right) \times \left(\dfrac{0.555 \text{ mole } H_2SO_4}{1 \text{ liter } H_2SO_4}\right)$

$\qquad\qquad \times \left(\dfrac{2 \text{ moles } CO_2}{1 \text{ mole } H_2SO_4}\right) \times \left(\dfrac{22.4 \text{ liters } CO_2}{1 \text{ mole } CO_2}\right) \times \left(\dfrac{1000 \text{ mL } CO_2}{1 \text{ liter } CO_2}\right)$

$\qquad\qquad\quad = 621.6$ mL CO_2 (calculator answer)
$\qquad\qquad\quad = 622$ mL CO_2 (correct answer)

c) g $NaHCO_3 \longrightarrow$ mole $NaHCO_3 \longrightarrow$ mole H_2SO_4

$\qquad$ 1.85 g $NaHCO_3 \times \left(\dfrac{1 \text{ mole } NaHCO_3}{84.0 \text{ g } NaHCO_3}\right) \times \left(\dfrac{1 \text{ mole } H_2SO_4}{2 \text{ moles } NaHCO_3}\right)$

$\qquad\qquad = 0.0110119$ mole H_2SO_4 (calculator answer)
$\qquad\qquad = 0.0110$ mole H_2SO_4 (correct answer)

$$M = \underline{\text{moles}} = \left(\frac{0.0110 \text{ mole } H_2SO_4}{32.7 \text{ mL } H_2SO_4 \text{ solution}}\right) \times \left(\frac{1000 \text{ mL}}{1 \text{ liter}}\right)$$

$$= 0.33639144 \text{ M } H_2SO_4 \text{ (calculator answer)}$$

$$= 0.336 \text{ M } H_2SO_4 \text{ (correct answer)}$$

11. Volume $AgNO_3$ solution $\longrightarrow$ moles $AgNO_3$ $\longrightarrow$ moles Ag_3PO_4

$\longrightarrow$ g Ag_3PO_4

$$25.0 \text{ mL } AgNO_3 \times \left(\frac{1 \text{ liter } AgNO_3}{1000 \text{ mL } AgNO_3}\right) \times \left(\frac{0.254 \text{ mole } AgNO_3}{1 \text{ liter } AgNO_3}\right)$$

$$\times \left(\frac{1 \text{ mole } Ag_3PO_4}{3 \text{ moles } AgNO_3}\right) \times \left(\frac{419 \text{ g } Ag_3PO_4}{1 \text{ mole } Ag_3PO_4}\right)$$

$$= 0.88688333 \text{ g } Ag_3PO_4 \text{ (calculator answer)}$$

$$= 0.887 \text{ g } Ag_3PO_4 \text{ (correct answer)}$$

ANSWERS TO MULTIPLE CHOICE EXERCISES

1. a

2. c

3. b

4. a

5. a

6. b

7. d

8. b

9. b

10. a

11. b

12. b

13. d

14. b

15. a

16. c

17. b

18. d

19. a

20. a

21. b

CHAPTER 14
Acids, Bases and Salts

REVIEW OF CHAPTER OBJECTIVES

1. State the definition of acids and bases using (1) the Arrhenius concept, (2) the Bronsted-Lowry concept (section 14.1).

    The Arrhenius concept of acids and bases defines them in terms of the species they form when dissolved in water. According to the Arrhenius concept, an acid is a substance that releases hydrogen ions (H^+) in water solution; a base is a substance that releases hydroxide ions (OH^-) in water solution.

    Arrhenius acids include the common acids HCl, HNO_3 and H_2SO_4 (just to list a few). When these substances are pure they are covalently bonded. They form hydrogen ions by reacting with the water they dissolve in.

    Arrhenius bases include NaOH and KOH, compounds which contain ions. The ions separate when the compounds dissolve in water. The Arrhenius definition of acids and bases is valid only in water solution.

    According to the Bronsted-Lowry concept, an acid is any substance that can donate a proton to another substance, and a base is any substance that can accept a proton from another substance. Another way to state this is to say that an acid is a proton donor and a base is a proton acceptor. Among the advantages of the Bronsted-Lowry concept of acids and bases is that it allows for acid-base reactions in nonaqueous solvents, and also in the gaseous state where there is no solvent present.

    All Arrhenius acids are Bronsted-Lowry acids, and all Arrhenius bases are Bronsted-Lowry bases, but not vice versa.

2. <u>Differentiate between, and give examples of, strong and</u>
   <u>weak Arrhenius acids and bases</u> (section 14.2).

   Strong Arrhenius acids are completely ionized in water
   solution. The common strong acids are HCl, HBr, HI,
   HNO_3, H_2SO_4 and $HClO_4$. Weak acids are only slightly
   ionized in water solution. The common weak acids are
   H_3PO_4, $HC_2H_3O_2$ and H_2CO_3 and other carbon-containing
   acids, HNO_2, H_2SO_3 and H_3BO_3. In water solution, strong
   acids exist entirely in the form of ions (no undissociated
   acid molecules are present in the water), but water
   solutions of weak acids contain most of the acid in
   molecular form rather than as ions.

   Strong Arrhenius bases are also completely ionized in
   water solution. Strong bases are limited to the
   hydroxides of Group IA and IIA. NaOH and KOH are the
   common strong bases found in the laboratory. LiOH, RbOH
   and CsOH are other strong Group IA bases. Of Group IIA
   hydroxides, only $Sr(OH)_2$ and $Ba(OH)_2$ are soluble, but
   the other Group IIA hydroxides, such as $Ca(OH)_2$, are also
   completely ionized although not very soluble in water.
   The only common weak base is NH_4OH, which more correctly
   should be referred to as "aqueous ammonia."

   You need to learn which acids and bases are strong, and
   which are weak. Such knowledge is a prerequisite for
   writing ionic equations later in this chapter. Also,
   remember that the terms "strong" or "weak" refer to the
   extent of dissociation of solute, not to the
   concentration of solution. A very dilute solution of HCl,
   such as 0.001 M HCl solution, still is a solution of a
   strong acid.

3. <u>Differentiate between, and give examples of, monoprotic,</u>
   <u>diprotic and triprotic acids, and write equations for the</u>
   <u>stepwise dissociation of diprotic and triprotic acids</u>
   (section 14.3).

   Acids may be classified according to the number of
   hydrogen ions they produce per molecule, on complete
   dissociation.

   A <u>monoprotic acid</u>, such as HCl, HNO_3 or $HC_2H_3O_2$, yields
   one hydrogen ion (H^+) per molecule on complete
   dissociation.

   A <u>diprotic acid</u> such as H_2SO_4 or H_2CO_3 yields two hydrogen ions (H^+) per molecule on complete dissociation.

The dissociation of a diprotic acid occurs in two steps, the second of which generally occurs to a lesser extent than the first.

$$H_2SO_4 = H^+ + HSO_4^-$$

$$HSO_4^- = H^+ + SO_4^{2-}$$

A <u>triprotic acid</u> such as H_3PO_4 yields three hydrogen ions (H^+) per molecule on complete dissociation. The dissociation of a triprotic acid also occurs stepwise.

$$H_3PO_4 = H^+ + H_2PO_4^-$$

$$H_2PO_4^- = H^+ + HPO_4^{2-}$$

$$HPO_4^{2-} = H^+ + PO_4^{3-}$$

Diprotic acids and triprotic acids together may be called polyprotic acids. A <u>polyprotic acid</u> is any acid capable of producing two or more hydrogen ions (H^+) per molecule on complete dissociation.

4. <u>State the meaning of the terms dilute and concentrated when applied to specific common acid and base stock solutions</u> (section 14.4).

Stock solutions of common acids and bases are found in almost all chemical laboratories. Traditionally the concentrations of such solutions are the same from laboratory to laboratory. Usually only the name of the acid or base and the term dilute (dil) or concentrated (con) are found on the containers for such solutions. It is assumed that the student or researcher knows what is implied by the designations dil and con in each specific case. Each specific case is a very important phrase in the last sentence since the meaning of the terms dil and con varies from acid to acid (and base to base) as shown in the following table.

Stock Laboratory Solutions

Acids	Concentrated	Dilute
HCl	12 M = 12 N	6 M = 6 N
HNO_3	16 M = 16 N	6 M = 6 N
H_2SO_4	18 M = 36 N	3 M = 6 N
$HC_2H_3O_2$	18 M = 18 N	6 M = 6 N
Bases		
NaOH		6 M = 6 N
aqueous ammonia (NH_4OH)	15 M = 15 N	6 M = 6 N

5. <u>Recognize compounds classified as salts</u> (section 14.5).

A salt is any ionic compound which contains a positive ion other than the hydrogen ion (H^+), and a negative ion other than the hydroxide ion (OH^-). Whenever an acid and a base react, water and a salt are produced. As an example:

$$KOH + HNO_3 \longrightarrow KNO_3 + H_2O$$

base   acid   salt

All common salts dissociate into ions in water solution. Even for salts that are only slightly soluble in water the part that does dissolve completely separates into ions. Therefore, there are no "weak" salts. The terms weak and strong are used to describe acids and bases, but not salts.

6. <u>Convert molecular equations into net ionic equations</u> (section 14.6).

So far, molecular equations have been used to describe chemical reactions. Molecular equations are equations which show the complete formulas of all reactants and products. Other types of equations exist. In an <u>ionic equation</u>, compounds completely dissociated in solution are written as ions, and undissociated compounds are written in molecular form. Weak acids and weak bases are shown in their predominant form, as molecular species.

Acids, Bases and Salts   287

A _net ionic equation_ is an ionic equation from which nonparticipating (spectator) ions have been eliminated. The examples below illustrate the difference between molecular, ionic and net ionic equations, for the reaction in which solutions of potassium chromate and lead(II) nitrate react to form the insoluble salt lead(II) chromate and the soluble salt potassium nitrate.

Molecular equation:

$$K_2CrO_{4(aq)} + Pb(NO_3)_{2(aq)} \longrightarrow PbCrO_{4(s)} + 2KNO_{3(aq)}$$

All but the $PbCrO_4$ are soluble salts, and will therefore be written when the molecular equation is converted to an ionic equation.

Ionic equation:

$$2K^+ + CrO_4^{2-} + Pb^{2+} + 2NO_3^- \longrightarrow PbCrO_4 + 2K^+ + 2NO_3^-$$

The potassium ions (K^+) and the nitrate ions (NO_3^-) appear on both sides of the equation, unchanged; they are spectator ions. In a net ionic equation such spectator ions are eliminated (cancelled out).

Net ionic equation:

$$Pb^{2+} + CrO_4^{2-} \longrightarrow PbCrO_4$$

This net ionic equation states that lead(II) ions react with chromate ions to form lead(II) chromate.

To convert molecular equations to net ionic equations, the steps followed are:

1. Make sure the molecular equation is balanced.

2. Change the molecular equation to an ionic equation by writing completely dissociated soluble compounds as separate ions. Soluble salts, strong acids and strong bases are all completely dissociated in solution.

   Weak acids and weak bases as well as insoluble salts, gases and covalent molecules such as water are left in molecular form.

3. Change the ionic equation to a net ionic equation by eliminating spectator ions.

4. Make sure the net ionic equation shows the smallest possible coefficients (divide all coefficients by any lowest common multiple).

Example: Write net ionic equations for the following reactions which occur in aqueous solution. (All of the equations are balanced.)

a) $Mg(NO_3)_2 + (NH_4)_2CO_3 \longrightarrow MgCO_3 + 2NH_4NO_3$

b) $2HCl + Ca(C_2H_3O_2)_2 \longrightarrow 2HC_2H_3O_2 + CaCl_2$

c) $H_2SO_4 + BaCl_2 \longrightarrow BaSO_4 + 2HCl$

Solution:

a) We must first decide which species are to be written in ionic form and which species are to be left in molecular form.

All four compounds in this particular equation are salts. Three of the four salts are soluble $(Mg(NO_3)_2, (NH_4)_2CO_3$ and $NH_4NO_3)$, and one is insoluble $(MgCO_3)$. The soluble salts are written in ionic form and the insoluble salt is left in molecular form:

$$Mg^{2+} + 2NO_3^- + 2NH_4^+ + CO_3^{2-} \longrightarrow MgCO_3 + 2NH_4^+ + 2NO_3^-$$

The NO_3^- and NH_4^+ ions are spectator ions and thus may be cancelled from the equation giving as the net ionic equation:

$$Mg^{2+} + CO_3^{2-} \longrightarrow MgCO_3$$

b) Two of the compounds in this equation are acids $(HCl$ and $HC_2H_3O_2)$ and two are salts $(Ca(C_2H_3O_2)_2$ and $CaCl_2)$.

HCl (a strong acid) is written in ionic form; $HC_2H_3O_2$ (a weak acid) is written in molecular form. Both of the salts are soluble and are thus written in ionic form.

$$2H^+ + 2Cl^- + Ca^{2+} + 2C_2H_3O_2^- \longrightarrow 2HC_2H_3O_2 + Ca^{2+} + 2Cl^-$$

The Ca^{2+} and Cl^- ions are spectator ions and may be cancelled, giving as the net ionic equation:

$$2H^+ + 2C_2H_3O_2^- \longrightarrow 2HC_2H_3O_2$$

The coefficients in this equation are all reducible (divisible by 2). Thus, the final net ionic equation is:

$$H^+ + C_2H_3O_2^- \longrightarrow HC_2H_3O_2$$

c) As in part b) this equation contains 2 acids (HCl and H_2SO_4) and 2 salts ($BaCl_2$ and $BaSO_4$).

Both of the acids are strong acids and are written in ionic form. $BaCl_2$ is a soluble salt (ionic form) and $BaSO_4$ is an insoluble salt (molecular form).

$$2H^+ + SO_4^{2-} + Ba^{2+} + 2Cl^- \longrightarrow BaSO_4 + 2H^+ + 2Cl^-$$

Cancelling out the spectator ions (H^+ and Cl^-) gives as a net ionic equation:

$$Ba^{2+} + SO_4^{2-} \longrightarrow BaSO_4$$

7. <u>Write equations for the reactions of acids with active metals, bases and carbonate and bicarbonate salts</u> (section 14.7).

In this section three general types of reactions that acids undergo are discussed. These three types of reactions are:

1. Acids react with active metals (those above hydrogen on the activity series) to produce hydrogen gas and a salt.

   Example: $Zn + 2HCl \longrightarrow ZnCl_2 + H_2$

2. Acids react with bases to produce a salt and water.

   Example: $HCl + KOH \longrightarrow KCl + H_2O$

3. Acids react with carbonates and bicarbonates to produce carbon dioxide (gas), a salt and water.

   Examples: $2HCl + Na_2CO_3 \longrightarrow 2NaCl + CO_2 + H_2O$

   $HCl + NaHCO_3 \longrightarrow NaCl + CO_2 + H_2O$

Example: Indicate what the products are in each of the following reactions by writing a balanced molecular equation for the reaction:

   a) $Mg + HC_2H_3O_2 \longrightarrow$?

b) $Ba(OH)_2 + H_2SO_4 \longrightarrow$?

c) $CaCO_3 + HNO_3 \longrightarrow$?

d) $Ca(HCO_3)_2 + HNO_3 \longrightarrow$?

Solution:

a) Magnesium is a metal above hydrogen in the activity series (see Table 14-6 on page 387 in the textbook). Therefore, it will react with acids to produce hydrogen gas (H_2).

$$Mg + 2HC_2H_3O_2 \longrightarrow Mg(C_2H_3O_2)_2 + H_2$$

b) The reaction of an acid and a base (neutralization) forms water plus a salt. The salt formed contains the negative ion from the acid and the positive ion from the base. H^+ ions from the acid and OH^- ions from the base produce the water.

$$H_2SO_4 + Ba(OH)_2 \longrightarrow BaSO_4 + 2H_2O$$

c) Acids react with carbonates to produce carbon dioxide, a salt and water.

$$CaCO_3 + 2HNO_3 \longrightarrow Ca(NO_3)_2 + H_2O + CO_2$$

d) Acids react with bicarbonates to produce carbon dioxide, a salt and water (the same products as for carbonates).

$$Ca(HCO_3)_2 + 2HNO_3 \longrightarrow Ca(NO_3)_2 + 2CO_2 + 2H_2O$$

8. Write equations for the reactions of salts in solution with selected metals, acids, bases and other salts (section 14.9).

Characteristic reactions for salts include the following four types of behavior:

1. Salts react with some metals to form another salt and another metal. This type of reaction will occur only if the metal going into solution is above the replaced metal in the activity series.

   Example: $Cu + 2AgNO_3 \longrightarrow Cu(NO_3)_2 + 2Ag$

2. Salts will react with acid solutions to form other acids and salts provided the new acid formed is a

Acids, Bases and Salts   291

weaker acid, the new salt is insoluble or a gaseous compound forms as one of the products.

Examples: $BaCl_2 + H_2SO_4 \longrightarrow BaSO_4 + 2HCl$

insoluble
salt

$$KC_2H_3O_2 + HCl \longrightarrow KCl + HC_2H_3O_2$$

weaker
acid

$$K_2CO_3 + 2HNO_3 \longrightarrow 2KNO_3 + H_2O + CO_2$$

gaseous
product

3. Salts will react with base solutions to form other bases and salts provided one of the products is a weaker base, an insoluble salt or a gaseous compound.

Example: $(NH_4)_2SO_4 + 2NaOH \longrightarrow Na_2SO_4 + 2NH_3 + 2H_2O$

gaseous
compound

4. Salts react with some solutions of other salts to form new salts. This type of reaction occurs only if an insoluble salt forms as one of the products.

Example: $Na_2S + CuCl_2 \longrightarrow 2NaCl + CuS$

insoluble
salt

Example: Indicate what the products are in each of the following reactions by writing a balanced molecular equation for the reaction.

a) $Zn + Pb(NO_3)_2 \longrightarrow$

b) $LiC_2H_3O_2 + HCl \longrightarrow$

c) $Al(NO_3)_3 + NaOH \longrightarrow$

d) $Pb(NO_3)_2 + KCl \longrightarrow$

Solution:

a) This reaction occurs because Zn is more active than Pb (Zn is above Pb on the activity series).

$$Zn + Pb(NO_3)_2 \longrightarrow Zn(NO_3)_2 + Pb$$

b) This reaction occurs because $HC_2H_3O_2$ is a weaker acid than HCl.

$$LiC_2H_3O_2 + HCl \longrightarrow HC_2H_3O_2 + LiCl$$

c) This reaction occurs because $Al(OH)_3$ is insoluble.

$$Al(NO_3)_3 + 3NaOH \longrightarrow 3NaNO_3 + Al(OH)_3$$

d) This reaction occurs because $PbCl_2$ is insoluble.

$$Pb(NO_3)_2 + 2KCl \longrightarrow PbCl_2 + 2KNO_3$$

9. <u>Give the value of $[H^+]$ and $[OH^-]$ in pure water at $25^\circ C$, and be able to calculate either the $[H^+]$ or the $[OH^-]$ in any water solution when the other is given</u> (section 14.10).

Water exists predominantly in the form of undissociated water molecules. However, a very few water molecules are ionized.

$$2H_2O = H_3O^+ + OH^-$$

or, more simply stated:

$$H_2O = H^+ + OH^-$$

In pure water, $[H^+] = [OH^-]$. At $25^\circ C$, both $[H^+]$ and $[OH^-] = 1.00 \times 10^{-7}$ M. This means that, in pure water, at $25^\circ C$ there are 2 H^+ ions and 2 OH^- ions for every billion undissociated water molecules.

At a given temperature, the product of the $[H^+]$ and $[OH^-]$ is constant.

At $25^\circ C$: $[H^+][OH^-] = (1.00 \times 10^{-7})(1.00 \times 10^{-7})$

$$= 1.00 \times 10^{-14}$$

This value is called the ion product of water. This ion product value applies to all water solutions at $25^\circ C$.

Example: Calculate the $[OH^-]$ in a solution at $25^\circ C$ which has a $[H^+] = 1.00 \times 10^{-3}$.

Solution:

$$[H^+][OH^-] = 1.00 \times 10^{-14}$$

$$[OH^-] = \frac{1.00 \times 10^{-14}}{[H^+]} = \frac{1.00 \times 10^{-14}}{1.00 \times 10^{-3}}$$

$$= 1 \times 10^{-11} \text{ (calculator answer)}$$

$$= 1.00 \times 10^{-11} \text{ (correct answer)}$$

Example: Calculate the $[H^+]$ in a solution at $25^\circ C$ where the $[OH^-] = 1.00 \times 10^{-12}$

Solution:

$$[H^+][OH^-] = 1.00 \times 10^{-14}$$

$$[H^+] = \frac{1.00 \times 10^{-14}}{[OH^-]} = \frac{1.00 \times 10^{-14}}{1.00 \times 10^{-12}}$$

$$= 1 \times 10^{-2} \text{ (calculator answer)}$$

$$= 1.00 \times 10^{-2} \text{ (correct answer)}$$

10. <u>Tell whether a solution is acidic or basic or neutral,
    given its $[H^+]$ or pH</u> (sections 14.10 and 14.11).

The relationship between the $[H^+]$ and $[OH^-]$ in an
aqueous solution is the basis for classifying the
solution as acidic, basic or neutral. All acidic
solutions have a higher $[H^+]$ than $[OH^-]$. All basic
solutions have a higher $[OH^-]$ than $[H^+]$. In a neutral
solution, the concentrations of both H^+ ions and OH^- ions
are equal.

In terms of pH a solution is acidic if its pH is less
than 7, basic if its pH is greater than 7 and neutral if
its pH is equal to 7.

Example: Indicate whether each of the following solutions
is "acidic" or "basic" or "neutral."

a) Human whole blood, pH = 7.4
b) Cow's milk, pH = 6.5

Solution:

a) Human whole blood, pH = 7.4, has a pH greater
than 7, so it is basic (though only slightly
basic).

294   Acids, Bases and Salts

b) Cow's milk, pH = 6.5, has a pH less than 7, so it is acidic.

11. Define pH, and calculate the pH of solutions with selected $[H^+]$ or $[OH^-]$ (section 14.11).

pH is a measure of the acidity of a solution. The pH scale runs from 0 – 14. When the $[H^+]$ is expressed in scientific notation, as 1.00×10^{-x}, then the pH = x.

A 0.0100 M HCl solution is completely ionized, so the $[H^+] = 1.00 \times 10^{-2}$ and the pH = 2.

A 0.0000100 M HCl solution has a $[H^+] = 1.00 \times 10^{-5}$, and pH = 5.

Example: State the $[H^+]$ in powers of ten, and the pH, for the following solutions:

a) 0.100 M HCl solution
b) 0.00100 M NaOH solution

Solution:

a) In 0.100 M HCl solution, since HCl is a strong acid (completely dissociated),

$$[H^+] = 1.00 \times 10^{-1}, \text{ and pH} = 1$$

b) In 0.00100 M NaOH solution, since NaOH is a strong base (completely ionized),

$$[OH^-] = 1.00 \times 10^{-3}$$

pH is based on the $[H^+]$, not on the $[OH^-]$. Therefore, the $[OH^-]$ is used to calculate the $[H^+]$ and then the pH is determined.

$$[H^+][OH^-] = 1.00 \times 10^{-14}$$

$$[H^+] = \frac{1.00 \times 10^{-14}}{[OH^-]} = \frac{1.00 \times 10^{-14}}{1.00 \times 10^{-3}}$$

$$= 1 \times 10^{-11} \text{ (calculator answer)}$$

$$= 1.00 \times 10^{-11} \text{ (correct answer)}$$

When the $[H^+]$ of a solution = 1.00×10^{-11}, the solution pH is 11.

12. <u>Understand the steps involved in the titration of an acid or base</u> (section 14.12).

An acid-base titration is a process in which an acid (or base) solution is gradually added to a base (or acid) solution until the solutes in the two solutions have completely reacted with each other.

In order to complete a titration successfully, we must be able to detect when the two solutes have completely reacted. An indicator is added to the solution being titrated for this purpose. An indicator is a compound that exhibits different colors in solutions of different pH. For example, the indicator phenolphthalein is colorless in solutions whose pH is less than about 8, and pink in solutions whose pH is more than about 8. So, as a basic solution is added to an acid solution which contains phenolphthalein, the solution would be seen to change from colorless to pink as the pH increased above about 8.

13. <u>Express the concentration of an acid or base solution in terms of normality (N)</u> (section 14.13).

Normality is the most convenient solution concentration unit to use for acid-base titrations. The reason for this is that it takes into account the fact that all acids and bases do not yield the same number of hydrogen or hydroxide ions per molecule on dissociation.

The definition of normality for use in dealing with acidic and basic solutions is:

$$N = \frac{\text{equivalents of acid or base}}{\text{L of solution}}$$

An equivalent of acid is the quantity of acid supplying one mole of H^+ ions on complete dissociation. An equivalent of base is the quantity of base that will react with one mole of H^+ ions.

Example: State the number of equivalents of acid or base in each of the following:

a) 1 mole H_2SO_4, b) 1 mole HCl, c) 1 mole NaOH

Solution:

a) One mole of H_2SO_4 yields two moles of H^+ upon dissociation:

$$H_2SO_4 \longrightarrow 2H^+ + SO_4^{2-}$$

Therefore,

$$1 \text{ mole } H_2SO_4 = 2 \text{ equivalents } H_2SO_4$$

b) Similarly, one mole of HCl will produce one mole of H^+:

$$HCl \longrightarrow H^+ + Cl^-$$

Therefore,

$$1 \text{ mole } HCl = 1 \text{ equivalent } HCl$$

c) The dissociation equation for NaOH is:

$$NaOH \longrightarrow Na^+ + OH^-$$

Therefore, 1 mole of NaOH gives up 1 mole of OH^- ion (which can react with 1 mole of H^+ ion). Therefore,

$$1 \text{ mole } NaOH = 1 \text{ equivalent } NaOH$$

Example: Calculate the normality (N) of a solution:

    a) containing 25.0 g H_2SO_4 in $\overline{300}$ mL solution

    b) labeled 0.272 M H_3PO_4

Solution:

a) For H_2SO_4, 2 equivalents are equal to 1 mole. Using dimensional analysis, the problem set-up is:

$$\left(\frac{25.0 \text{ g } H_2SO_4}{300 \text{ mL}}\right) \times \left(\frac{1000 \text{ mL}}{1 \text{ liter}}\right) \times \left(\frac{1 \text{ mole } H_2SO_4}{98.1 \text{ g } H_2SO_4}\right)$$

$$\times \left(\frac{2 \text{ equivalents } H_2SO_4}{1 \text{ mole } H_2SO_4}\right) = 1.6989465 \text{ N (calculator answer)}$$

$$= 1.70 \text{ N (correct answer)}$$

b) For H_3PO_4, 3 equivalents are equal to 1 mole. Therefore,

$$N = \frac{\text{equivalents}}{\text{liter}} = \left(\frac{0.272 \ \text{moles } H_3PO_4}{\text{liter}}\right)$$

$$x \left(\frac{3 \ \text{equivalents } H_3PO_4}{1 \ \text{mole } H_3PO_4}\right) = 0.816 \ N \ H_3PO_4 \ \text{solution}$$
(calculator and correct answer)

14. <u>Calculate the volume (or concentration) of an acidic or basic solution, given its concentration (or volume) and the volume and concentration of a second solution with which it is titrated (section 14.13).</u>

In an acid–base titration the concentration (normality) of one of the solutions can be calculated when the concentration (normality) of the other is known and the volumes of both solutions are also known. The equation for such a calculation is:

$$(N_{acid}) \ x \ (V_{acid}) = (N_{base}) \ x \ (V_{base})$$

Example: What is the normality of an NaOH solution if 30.0 mL of this NaOH solution neutralizes 25.0 mL of 0.572 N H_3PO_4 solution according to the equation:

$$H_3PO_4 + 3NaOH \longrightarrow 3H_2O + Na_3PO_4$$

Solution:

$$N_a V_a = N_b V_b$$

$$N_b = \frac{N_a V_a}{V_b} = \frac{25.0 \ \text{mL} \ x \ 0.572 \ N}{30.0 \ \text{mL}} = 0.47666667 \ N \ NaOH$$
(calculator answer)
$$= 0.477 \ N \ NaOH$$
(correct answer)

Example: How many mL of 0.275 N H_2SO_4 solution are needed to neutralize 22.5 mL of 0.352 N NaOH solution according to the equation:

$$H_2SO_4 + 2NaOH \longrightarrow 2H_2O + Na_2SO_4$$

Solution:

$$N_a V_a = N_b V_b$$

$$V_a = \frac{V_b N_b}{N_a} = \frac{22.5 \ \text{mL} \ x \ 0.352 \ N}{0.275 \ N} =$$

= 28.8 mL H_2SO_4 solution (calculator and correct answer)

PROBLEM SET

1. Define acids and bases according to the Arrhenius concept.

2. Define acids and bases according to the Bronsted–Lowry concept.

3. a) Explain the difference between strong acids and weak acids.

   b) Explain the difference between strong bases and weak bases.

   c) Why are there no weak salts?

4. Explain why a 0.1 M HCl solution has a lower pH than a 0.1 M $HC_2H_3O_2$ solution.

5. Write net ionic equations for the following reactions. The equations given are not balanced:

   a) H_2SO_4 + NaOH = H_2O + Na_2SO_4
   b) NaOH + $Fe(NO_3)_3$ = $Fe(OH)_3$ + $NaNO_3$
   c) HNO_3 + K_2CO_3 = KNO_3 + CO_2 + H_2O
   d) Zn + $Cu(NO_3)_2$ = $Zn(NO_3)_2$ + Cu

6. A solution contains 20.0 g of $HC_2H_3O_2$ in $\overline{200}$ mL of solution. Calculate the molarity (M) and normality (N) of this solution.

7. How many grams of $Sr(OH)_2$ are in 25.0 mL of 0.0520 N $Sr(OH)_2$ solution?

8. If 25.0 mL of 0.250 N H_2SO_4 solution is needed to neutralize 32.5 mL of NaOH solution, what is the normality (N) of the NaOH solution?

9. How many mL of 0.254 N H_2SO_4 solution are needed to neutralize 15.2 mL of 0.350 N KOH solution?

10. If 25.0 mL of a H_3PO_4 solution just neutralizes 30.0 mL of 0.585 N NaOH solution, how many:

a) equivalents of H_3PO_4 are present?

b) moles of H_3PO_4 are present?

c) grams of H_3PO_4 are present?

MULTIPLE CHOICE EXERCISES

1. According to the Bronsted-Lowry theory, an acid is:

   a) a proton donor
   b) a proton acceptor
   c) a hydronium ion donor
   d) a hydronium acceptor

2. According to Arrhenius acid-base theory, the properties of acids are the properties of the:

   a) H^+ ion

   b) OH^- ion

   c) H_2O molecule

   d) Cl^- ion

3. A strong acid is one which:

   a) tastes sour
   b) reacts with bases
   c) ionizes nearly 100% in solution
   d) has large muscles

4. Which of the following is not a strong acid?

   a) HCl
   b) H_3BO_3
   c) H_2SO_4
   d) HNO_3

5. Which of the following is a species formed in the second step of the dissociation of H_3PO_3?

   a) $H_2PO_3^-$

   b) HPO_3^{2-}

   c) PO_3^{2-}

   d) $2H^+$

6. Which of the following statements concerning acid and base stock solutions is incorrect?

   a) dilute HCl is 6M

   b) dilute H_2SO_4 is 6M

   c) concentrated HNO_3 is 16 M

   d) dilute NaOH is 6M

7. Which of the following is a salt?

   a) H_2SO_4

   b) $BaSO_4$

   c) $Ba(OH)_2$

   d) NaOH

8. The net ionic equation for the reaction:

$$NaCl + AgNO_3 \longrightarrow AgCl + NaNO_3$$

   is:

   a) $Ag^+ + Cl^- \longrightarrow AgCl$

   b) $Na^+ + NO_3^- \longrightarrow NaNO_3$

   c) $Na^+ + AgNO_3 \longrightarrow Ag^+ + NaNO_3$

   d) $NaCl + NO_3^- \longrightarrow NaNO_3 + Cl^-$

9. The net ionic equation for the reaction:

$$H_2S + Cu(NO_3)_2 \longrightarrow 2HNO_3 + CuS$$

   is:

   a) $H_2S + Cu^{2+} \longrightarrow 2H^+ + CuS$

   b) $S^{2-} + Cu^{2+} \longrightarrow CuS$

c) $S^{2-} + Cu(NO_3)_2 \longrightarrow 2NO_3^- + CuS$

d) $2H^+ + Cu^{2+} + 2NO_3^- \longrightarrow Cu + 2HNO_3$

10. Acids react with active metals to produce:

    a) a salt and water
    b) hydrogen gas and a salt
    c) carbon dioxide, a salt and water
    d) inactive metals

11. Which of the following is a neutralization reaction:

    a) $CaCO_3 \longrightarrow CaO + CO_2$
    b) $Zn + H_2SO_4 \longrightarrow ZnSO_4 + H_2$
    c) $Ba(OH)_2 + 2HCl \longrightarrow BaCl_2 + 2H_2O$
    d) $NaCl + AgNO_3 \longrightarrow AgCl + NaNO_3$

12. Two different salt solutions will react when mixed only if which of the following conditions is met?

    a) an insoluble salt is produced
    b) a weak acid is produced
    c) a strong base is produced
    d) a gaseous compound is produced

13. In pure water at $25^{\circ}C$ the hydrogen ion concentration, in moles per liter, is:

    a) zero
    b) 1×10^{-7}
    c) 1×10^{-14}
    d) 0.001

14. A solution has $[H^+] = 1 \times 10^{-2}$. This solution is best described as:

    a) strongly basic
    b) neutral
    c) weakly acidic
    d) strongly acidic

15. Which of the following solutions is basic?

    a) $[H^+] = 1 \times 10^{-3}$ c) $[H^+] = 1 \times 10^{-7}$
    b) $[H^+] = 1 \times 10^{-5}$ d) $[H^+] = 1 \times 10^{-9}$

16. The pH of a solution is 8. This means the hydrogen ion concentration is:

   a) 8 molar
   b) 1×10^8 molar
   c) 1×10^{-8} molar
   d) 8×10^{-1} molar

17. In a titration an indicator is used to:

   a) neutralize the acid present
   b) neutralize the base present
   c) make the endpoint of the titration visible
   d) make the acid visible

18. If 4 equivalents of an acid are dissolved in 500 mL of solution, the concentration is:

   a) 1.25 N
   b) 2 N
   c) 4 N
   d) 8 N

19. A solution of H_3PO_4 which is 0.90 N would also be:

   a) 0.30 M
   b) 0.45 M
   c) 0.90 M
   d) 2.70 M

20. If 25.0 mL of 0.100 N acid is neutralized by 20.0 mL of base, what is the concentration of the base?

   a) 0.080 N
   b) 0.100 N
   c) 0.125 N
   d) 0.150 N

SOLUTIONS TO PROBLEMS SET

1. Acids produce hydrogen ions in water and bases produce hydroxide ions in water.

2. Acids are proton donors and bases are proton acceptors.

3. a) Strong acids are completely dissociated in water
   solution, while weak acids are only partly dissociated
   in water solution.

   b) Strong bases are completely dissociated in water
   solution, while weak bases are only partly dissociated
   in water solution.

   c) Pure salts exist as ions, so any salt that dissolves
   in water is present in the form of ions (there are no
   salt "molecules," in the pure form or in water
   solution).

4. 0.1 M HCl solution has a hydrogen ion concentration of
   0.1 M because HCl is a strong acid (it is completely
   dissociated in water solution). 0.1 M $HC_2H_3O_2$ solution
   is only partly dissociated (since this acid is a weak
   acid) and so the hydrogen ion concentration is less than
   0.1 M. The pH of the HCl solution is lower; the greater
   the hydrogen ion concentration, the lower the pH.

5. a) $H_2SO_4 + 2NaOH \longrightarrow 2H_2O + Na_2SO_4$

   $2H^+ + \cancel{SO_4^{2-}} + 2\cancel{Na^+} + 2\ OH^- \longrightarrow 2H_2O + 2\cancel{Na^+} + \cancel{SO_4^{2-}}$

   $2H^+ + 2\ OH^- \longrightarrow 2H_2O$

   $H^+ + OH^- \longrightarrow H_2O$

   b) $3NaOH + Fe(NO_3)_3 \longrightarrow Fe(OH)_3 + 3NaNO_3$

   $3\cancel{Na^+} + 3\ OH^- + Fe^{3+} + 3\cancel{NO_3^-} \longrightarrow Fe(OH)_3 + 3\cancel{Na^+} + 3\cancel{NO_3^-}$

   $Fe^{3+} + 3\ OH^- \longrightarrow Fe(OH)_3$

   c) $2HNO_3 + K_2CO_3 \longrightarrow 2KNO_3 + CO_2 + H_2O$

   $2H^+ + 2\cancel{NO_3^-} + 2\cancel{K^+} + CO_3^{2-} \longrightarrow 2\cancel{K^+} + 2\cancel{NO_3^-} + CO_2 + H_2O$

   $2H^+ + CO_3^{2-} \longrightarrow CO_2 + H_2O$

   d) $Zn + Cu(NO_3)_2 \longrightarrow Zn(NO_3)_2 + Cu$

   $Zn + Cu^{2+} + 2\cancel{NO_3^-} \longrightarrow Zn^{2+} + 2\cancel{NO_3^-} + Cu$

   $Zn + Cu^{2+} \longrightarrow Zn^{2+} + Cu$

6. One equivalent $HC_2H_3O_2$ = 1 mole $HC_2H_3O_2$

$$M = \frac{\text{moles}}{\text{liter}} = \left(\frac{20.0\ \cancel{g}}{200\ \cancel{mL}}\right) \times \left(\frac{1\ \text{mole}}{60.0\ \cancel{g}}\right) \times \left(\frac{1000\ \cancel{mL}}{1\ \text{liter}}\right) = 1.6666667\ M \text{ (calculator answer)}$$

$$= 1.67\ M \text{ (correct answer)}$$

7. $$25.0\ \cancel{mL} \times \left(\frac{1\ \cancel{L}}{1000\ \cancel{mL}}\right) \times \left(\frac{0.0520\ \cancel{equiv}}{1\ \cancel{L}}\right) \times \left(\frac{1\ \cancel{mole}}{2\ \cancel{equiv}}\right) \times \left(\frac{121.6\ g}{1\ \cancel{mole}}\right)$$

$$= 0.07904\ g \text{ (calculator answer)}$$

$$= 0.0790\ g \text{ (correct answer)}$$

8. $$N_a V_a = N_b V_b$$

$$N_b = \frac{N_a V_a}{V_b} = \frac{25.0\ \cancel{mL} \times 0.250\ N}{32.5\ \cancel{mL}} = 0.19230769\ N \text{ (calculator answer)}$$

$$= 0.192\ N \text{ (correct answer)}$$

9. $$N_a V_a = N_b V_b$$

$$V_a = \frac{N_b V_b}{N_a} = \frac{0.350\ \cancel{N} \times 15.2\ \text{mL}}{0.254\ \cancel{N}} = 20.944882\ \text{mL} \text{ (calculator answer)}$$

$$= 20.9\ \text{mL (correct answer)}$$

10. a) $$N_a V_a = N_b V_b$$

$$N_a = \frac{N_b V_b}{V_a} = \frac{0.585\ N \times 30.0\ \cancel{mL}}{25.0\ \cancel{mL}} = 0.702\ N \text{ (calculator and correct answer)}$$

$$25.0\ \cancel{mL} \times \left(\frac{1\ \cancel{liter}}{1000\ \cancel{mL}}\right) \times \left(\frac{0.702\ \text{equivalents}}{1\ \cancel{liter}}\right)$$

$$= 0.01755\ \text{equiv (calculator answer)}$$
$$= 0.0176\ \text{equiv (correct answer)}$$

b) $$0.0176\ \cancel{equiv\ H_3PO_4} \times \left(\frac{1\ \text{mole } H_3PO_4}{3\ \cancel{equiv.\ H_3PO_4}}\right)$$

$$= 0.00586667\ \text{moles } H_3PO_4 \text{ (calculator answer)}$$

$$= 0.00587\ \text{moles } H_3PO_4 \text{ (correct answer)}$$

c) $$0.00587\ \cancel{moles\ H_3PO_4} \times \left(\frac{98.1\ g\ H_3PO_4}{1\ \cancel{mole\ H_3PO_4}}\right)$$

$$= 0.575847\ g\ H_3PO_4 \text{ (calculator answer)}$$

$$= 0.576\ g\ H_3PO_4 \text{ (correct answer)}$$

ANSWERS TO MULTIPLE CHOICE EXERCISES

1.	a	11.	c
2.	b	12.	a
3.	c	13.	b
4.	b	14.	d
5.	b	15.	d
6.	b	16.	c
7.	b	17.	c
8.	a	18.	d
9.	a	19.	a
10.	b	20.	c

CHAPTER 15
Oxidation and Reduction

REVIEW OF CHAPTER OBJECTIVES

1. <u>Define the terms oxidation, reduction, oxidizing agent and reducing agent in terms of loss and gain of electrons</u> (section 15.1).

Historically, the term oxidation meant reaction with oxygen and the term reduction meant removal of oxygen from a compound. Current definitions are much broader, and allow for use of these two terms even when no oxygen is involved.

The modern definitions are:

Oxidation is the process in which a substance loses electrons in a chemical reaction.

Reduction is the process in which a substance gains electrons in a chemical reaction.

Oxidation and reduction must occur together. Any reaction involving oxidation and reduction is called an oxidation-reduction reaction, or redox reaction (for short), and involves a transfer of electrons.

Since oxidation and reduction both occur in a chemical reaction we can also refer to the reactants as the oxidizing agent and the reducing agent.

The oxidizing agent causes oxidation by accepting electrons from the other reactant. This means that the oxidizing agent is itself reduced.

The reducing agent causes reduction by providing electrons for the other reactant to accept.

In other words, the oxidizing agent is the substance reduced and the reducing agent is the substance oxidized.

2. Determine the oxidation number of an element in a
   molecule or an ion (section 15.2).

   Oxidation numbers are used to determine whether oxidation
   or reduction occurs in a chemical reaction. Oxidation
   number is the charge that an atom appears to have when
   the electrons in each bond are assigned to the more
   electronegative of the two atoms involved in the bond.

   In an HF molecule: H$\overset{x}{.}$F:

   the shared electron pair (between hydrogen and fluorine)
   is assigned to F, the more electronegative element of the
   two. Therefore, the oxidation number of H is +1, and the
   oxidation number of F is -1.

                              H
                              x.
   In CH$_4$ H$\overset{x}{.}$C$\overset{x}{.}$H
                              .x
                              H

   each shared electron pair (between C and H) is assigned
   to the C (the more electronegative atom), so the
   oxidation number of each H is +1 and the oxidation number
   of C is -4.

   In the Cl$_2$ molecule,   :Cl$\overset{xx}{x}$Cl$\overset{x}{x}$:

   the electrons are shared equally (since identical atoms
   are of equal electronegativity) and neither atom is
   considered to have gained or lost electrons. Each
   chlorine atom, in Cl$_2$, has an oxidation number of zero.

   An alternative procedure for determining oxidation
   numbers, which does not require drawing electron-dot
   structures, uses a set of operational rules which are as
   follows:

   Rule 1:   The oxidation number of an element in its
             elemental state is zero.

   Rule 2:   The oxidation number of any monoatomic ion is
             equal to the charge on the ion.

   Rule 3:   The oxidation numbers of Group IA and IIA
             elements, in their compounds, are always +1 and
             +2, respectively.

   Rule 4:   The oxidation number of fluorine (in compounds)
             is always -1, and for the other Group VIIA

elements (Cl, Br, I) it is usually −1 (except
when they are bonded to more electronegative
elements).

Rule 5: The oxidation number of oxygen in compounds is
−2, except when oxygen is bonded to the more
electronegative element fluorine, or in
peroxides (where the oxidation number of oxygen
is −1).

Rule 6: The oxidation number of hydrogen in compounds is
+1, except in metal hydrides.

Rule 7: The algebraic sum of the oxidation numbers of
all atoms in a molecule must be zero.

Rule 8: The algebraic sum of the oxidation numbers of
all atoms in a polyatomic ion equals the charge
on the ion.

Example: Determine the oxidation number of each element:

a) K_2CrO_4, b) $KClO$, c) $KClO_2$, d) $NaClO_3$,
e) ClO_4^-

Solution:

The oxidation number of oxygen in each compound is
−2. The oxidation number of the group IA elements
(in a compound) is +1. The oxidation number of
hydrogen in its compounds is +1.

a) K_2CrO_4 b) $KClO$ c) $KClO_2$ d) $NaClO_3$
  +1+6−2   +1+1−2   +1+3−2   +1+5−2

e) ClO_4^-
  +7−2

Oxidation number is always specified per atom.

3. <u>Define the terms oxidation, reduction, oxidizing agent
and reducing agent in terms of increase and decrease in
oxidation number</u> (section 15.2).

The oxidizing agent in a redox reaction is the substance
containing the atom that decreases in oxidation number.

The reducing agent in a redox reaction is the substance
containing the atom that increases in oxidation number.

Oxidation and Reduction   309

Since the oxidizing agent is the substance reduced in a reaction, reduction involves a decrease in oxidation number. Since the reducing agent is the substance oxidized in a reaction, oxidation involves an increase in oxidation number.

4. <u>Identify in a redox reaction the substance oxidized and and substance reduced, the oxidizing agent and the reducing agent</u> (section 15.2).

> *Example:* Determine oxidation numbers for each atom in the following reaction and identify the oxidizing agent and the reducing agent.

Solution:

$$MnO_2 + 4HCl = MnCl_2 + Cl_2 + 2H_2O$$
$$+4-2 \quad +1-1 \quad +2-1 \quad 0 \quad +1-2$$

The oxidizing agent contains the atom decreasing in oxidation number. The oxidation number of Mn has decreased from +4 to +2. Therefore, the reactant that contains Mn, MnO_2, is the oxidizing agent.

The reducing agent contains the atom increasing in oxidation number. The oxidation number of Cl has increased from -1 to 0. Therefore, the reactant that contains Cl, HCl, is the reducing agent.

The substance reduced is the oxidizing agent, MnO_2. The substance oxidized is the reducing agent, HCl.

5. <u>Classify reactions as either redox reactions or metathetical reactions</u> (section 15.3).

A classification system based on oxidation number change groups chemical reactions into two types: redox (oxidation-reduction) reactions and metathetical reactions.

A redox reaction is one in which oxidation numbers change.

Example:
$$Mg + ZnSO_4 = Zn + MgSO_4$$
$$0 \quad +2+6-2 \quad 0 \quad +2+6-2$$

A metathetical reaction is one in which there is no oxidation number change.

Example: $AgNO_3$ + KCl = $AgCl$ + KNO_3

$+1+5-2$ $+1-1$ $+1-1$ $+1+5-2$

Example: Classify the following reactions as either redox or metathetical:

a) Na_2CO_3 + $2HCl$ = $2NaCl$ + H_2O + CO_2

b) Cu + $4HNO_3$ = $Cu(NO_3)_2$ + $2NO_2$ + $2H_2O$

Solution:

a) Assigning oxidation numbers we get:

Na_2CO_3 + $2HCl$ = $2NaCl$ + H_2O + CO_2

$+1+4-2$ $+1-1$ $+1-1$ $+1-2$ $+4-2$

There are no changes in oxidation number; this is a metathetical equation.

b) Cu + $4HNO_3$ = $Cu(NO_3)_2$ + $2NO_2$ + $2H_2O$

0 $+1+5-2$ $+2+5-2$ $+4-2$ $+1-2$

Cu and N both undergo oxidation number changes; this is a redox equation.

6. <u>Classify reactions as either synthesis, decomposition, single replacement or double replacement reactions</u> (section 15.3).

This classification system recognizes 4 types of reactions:

1. A <u>synthesis reaction</u>, in which a single product is produced from two (or more) reactants. Synthesis reactions include:

Zn + S = ZnS

C + O_2 = CO_2

CO_2 + H_2O = H_2CO_3

2. A <u>decomposition reaction</u>, in which a single reactant decomposes into two (or more) simpler substances. Decomposition reactions include:

$2HgO$ = $2Hg$ + O_2

$$Na_2CO_3 = Na_2O + CO_2$$

3. A _single replacement reaction_, in which an element replaces another element from its compound. Single replacement reactions include:

$$Mg + 2HCl = MgCl_2 + H_2$$

$$Cu + 2AgNO_3 = Cu(NO_3)_2 + 2Ag$$

$$Cl_2 + 2KBr = Br_2 + 2KCl$$

4. A _double replacement reaction_, in which two compounds exchange parts of the compounds with each other and form two different compounds. Double replacement reactions include:

$$HNO_3 + KOH = KNO_3 + H_2O$$

$$H_2S + Cu(NO_3)_2 = 2HNO_3 + CuS$$

7. <u>Balance a redox equation in neutral, acidic or basic solution, using the oxidation number method of balancing</u> (section 15.5).

Redox reactions are often complicated, and difficult to balance by inspection because some of the coefficients may be large. "Bookkeeping" on oxidation number changes helps balance these equations, because the number of electrons lost and gained in a redox reaction must be equal.

Two systematic approaches for balancing redox reactions are the oxidation number method and the ion-electron half-reaction method. In this section we will use the oxidation number method, which follows these steps:

1) Assign oxidation numbers to all atoms, and determine which atoms change oxidation numbers.

2) Determine the change in oxidation number per atom for elements changing oxidation number.

3) When more than one atom of an element which changes oxidation number is present in a formula, on either side of the equation, determine the change in oxidation number per formula unit.

4) Determine the multiplying factors that make the increase and decrease in oxidation numbers equal.

5) Use the multiplying factors (in step 4) as coefficients in the equation.

6) Balance all other atoms in the equation (except for hydrogen and oxygen) by inspection.

7) In net ionic equations, balance the charge (it must be the same on both sides) by adding H^+ (if reaction occurs in acid solution) or OH^- (if reaction occurs in basic solution).

8) Balance the oxygen atoms by adding H_2O as needed to a net ionic equation.

9) Check that the hydrogen atoms are balanced. If they are not, look for any mistakes that were made in earlier steps.

The examples below illustrate use of the oxidation number method of balancing redox equations.

Example: Balance the following equation using the oxidation-number method of balancing:

$$H_2SO_3 + HIO_3 = H_2SO_4 + HI$$

Solution:

Step 1:   Assign oxidation numbers to all atoms.

$$\begin{array}{cccc} H_2SO_3 & + \; HIO_3 & = \; H_2SO_4 & + \; HI \\ {\scriptstyle +1+4-2} & {\scriptstyle +1+5-2} & {\scriptstyle +1+6-2} & {\scriptstyle +1-1} \end{array}$$

S and I are the elements changing oxidation number.

Step 2:   Determine the change in oxidation number per atom.

Step 3:   This step doesn't apply as only one atom of each element that changes oxidation number is present in each formula unit.

Step 4. To make the number of electrons lost equal the number of electrons gained, the change involving S must be multiplied by 3.

$$\overset{+5 \qquad (-6) \qquad\quad -1}{\underset{+4 \qquad 3(+2) \qquad +6}{H_2SO_3 + HIO_3 = H_2SO_4 + HI}}$$

Step 5. Place the 3 as a coefficient in the equation.

$$3H_2SO_3 + HIO_3 = 3H_2SO_4 + HI$$

This equation is now balanced; steps 6 through 9 are not needed. Each side shows 3S, 1I, 7H and 12 O atoms.

Example. Balance the following equation using the oxidation-number method of balancing:

$$HNO_3 + HI = NO + I_2 + H_2O$$

Solution:

Step 1. Assign oxidation numbers for all atoms.

$$HNO_3 + HI = NO + I_2 + H_2O$$
$$+1+5-2 \quad +1-1 \quad +2-2 \quad 0 \quad +1-2$$

Step 2. Determine the change in oxidation number per atom.

$$\overset{-1 \qquad (+1) \qquad 0}{\underset{+5 \qquad (-3) \qquad +2}{HNO_3 + HI = NO + I_2 + H_2O}}$$

Step 3. Determine the change in oxidation number per formula unit.

$$\overset{2(+1)}{\underset{(-3)}{HNO_3 + HI = NO + I_2 + H_2O}}$$

Step 4. Determine numbers which will make loss and gain in oxidation number the same.

$$\overset{\overbrace{\qquad 3[2(+1)] \qquad}}{HNO_3 \; + \; HI \; = \; NO \; + \; I_2 \; + \; H_2O}$$
$$\underbrace{\qquad\qquad\qquad}_{2(-3)}$$

Step 5. Use the numbers from step 4 as coefficients in the equation.

$$2HNO_3 \; + \; 6HI \; = \; 2NO \; + \; 3I_2 \; + \; H_2O$$

Step 6. and beyond: Balance the hydrogen and oxygen.

$$2HNO_3 \; + \; 6HI \; = \; 2NO \; + \; 3I_2 \; + \; 4H_2O$$

Example: Balance the following net ionic equation using the oxidation-number method of balancing:

$$Fe^{2+} \; + \; MnO_4^{-} \; = \; Mn^{2+} \; + \; Fe^{3+}$$

(occurs in acidic solution)

Step 1. $Fe^{2+} \; + \; MnO_4^{-} \; = \; Mn^{2+} \; + \; Fe^{3+}$
$\quad\quad\;\; +2 \quad\;\; +7-2 \quad\quad\;\; +2 \quad\quad +3$

Fe and Mn undergo a change in oxidation number.

Steps 2, 3 and 4.

$$\overset{\overbrace{+7 \quad\; (-5) \quad\; +2}}{Fe^{2+} \; + \; MnO_4^{-} \; = \; Mn^{2+} \; + \; Fe^{3+}}$$
$$\underbrace{\; +2 \qquad\quad 5(+1) \qquad\qquad\quad +3 \;}$$

Step 5. Place these numbers in the equation.

$$5Fe^{2+} \; + \; MnO_4^{-} \; = \; Mn^{2+} \; + \; 5Fe^{3+}$$

Step 6. Does not apply.

Step 7. The total charge shown on each side of a net ionic equation must be the same. Since this reaction occurs in acidic solution, hydrogen ions are added (as needed) to balance the charge.

The left side shows a charge of +9 and the right side shows a charge of +17. By adding 8 H^+ to the left side the charge will be the same (+17) on both sides.

$$5Fe^{2+} + MnO_4^- + 8H^+ = Mn^{2+} + 5Fe^{3+}$$

Step 8.   Add water molecules to balance the oxygen.

$$5Fe^{2+} + MnO_4^- + 8H^+ = Mn^{2+} + 5Fe^{3+} + 4H_2O$$

Step 9.   The equation is balanced. Both sides show a charge of +17: 5 Fe, 1 Mn, 4 O and 8 H atoms are present on each side.

8.   <u>Balance a redox equation in neutral, acidic or basic</u> <u>solution using the ion-electron method of balancing</u> (section 15.6).

In the ion-electron method of balancing redox equations, two separate partial equations (called half-reactions) are written. A balanced oxidation half-reaction and a balanced reduction half-reaction are written and then added to get the desired balanced redox equation. The steps for this method are:

Step 1.   Write the equation in net ionic form and determine which substances are oxidized and reduced by assigning oxidation numbers to each atom or ion.

Step 2.   Write two separate partial equations:

a) An oxidation half-reaction involving the formula of the reactant oxidized and its product, and

b) a reduction half-reaction involving the formula of the reactant reduced and its product.

Step 3.   Balance each half-reaction with respect to the element oxidized or reduced, using appropriate coefficients.

Step 4.   Balance each half-reaction with respect to all other elements except hydrogen and oxygen.

Step 5.   Balance each half-reaction with respect to oxygen. In acidic or neutral solutions H_2O is added to the side deficient in oxygen. In basic solution, for every deficient oxygen, add two

OH^- ions to the side deficient in oxygen and one H_2O molecule to the other side of the equation.

Step 6. Balance each half-reaction with respect to hydrogen. In acidic solution, H^+ is added to the hydrogen deficient side. In basic solution, for every deficient hydrogen, add one H_2O molecule to the side deficient in H and then one hydroxide ion to the other side for every H_2O molecule added.

Step 7. Balance each half-reaction with respect to charge by adding electrons to the side with the more positive charge.

Step 8. Multiply each half-reaction by the lowest number that makes the number of electrons lost equal the number of electrons gained.

Step 9. Add the two half-reactions.

Step 10. Cancel out the electrons (which must be equal on both sides) and any other species common to both sides.

Step 11. Check that the equation coefficients are in the lowest possible ratio.

Example: Balance the following net ionic equation which occurs in acidic solution, using the ion-electron method of balancing:

$$IO_3^- + SO_2 = I_2 + SO_4^{2-}$$

Solution:

   Step 1. Assign an oxidation number to each atom.

$$IO_3^- + SO_2 = I_2 + SO_4^{2-}$$
$$+5 \quad\ +4 \qquad 0 \quad\ +6$$

   S is oxidized; it increases in oxidation number from +4 to +6. I is reduced; it decreases in oxidation number from +5 to 0.

   Step 2: The skeleton half reactions are:

   Oxidation $\qquad SO_2 = SO_4^{2-}$

Reduction $\quad IO_3^- = I_2$

Step 3. Balance the number of atoms for the elements oxidized and reduced.

Oxidation $\quad SO_2 = SO_4^{2-}$

Reduction $\quad 2IO_3^- = I_2$

Step 4. There are no other elements besides oxygen which need balancing.

Step 5. Since the reaction occurs in acid solution, oxygen is balanced by adding water to the oxygen deficient side.

Oxidation $\quad SO_2 + 2H_2O = SO_4^{2-}$

Reduction $\quad 2IO_3^- = I_2 + 6H_2O$

Step 6. Balance hydrogen by adding H^+ (as needed)

Oxidation $\quad SO_2 + 2H_2O = SO_4^{2-} + 4H^+$

Reduction $\quad 2IO_3^- + 12H^+ = I_2 + 6H_2O$

Step 7. Add electrons to balance each half-reaction with respect to charge.

Oxidation $\quad SO_2 + 2H_2O = SO_4^{2-} + 4H^+ + 2e^-$

Reduction $\quad 2IO_3^- + 12H^+ + 10e^- = I_2 + 6H_2O$

Step 8. Multiply each half-reaction by numbers which make the number of electrons lost equal the number of electrons gained.

Oxidation

$$5(SO_2 + 2H_2O = SO_4^{2-} + 4H^+ + 2e^-)$$

Reduction

$$2IO_3^- + 12H^+ + 10e^- = I_2 + 6H_2O$$

Step 9. Add the half-reactions, cancelling the electrons.

Oxidation $\quad 5SO_2 + 10H_2O = 5SO_4^{2-} + 20\ H^+ + 10e^-$

Reduction $\quad 2IO_3^- + 12H^+ + 10e^- = I_2 + 6H_2O$

$$5SO_2 + 2IO_3^- + 10H_2O + 12H^+ + \cancel{10e^-} =$$
$$5SO_4^{2-} + I_2 + 6H_2O + 20\ H^+ + \cancel{10e^-}$$

Step 10. Cancel any species present on both sides; in this case subtracting $6H_2O$ and $12H^+$ from both sides.

$$5SO_2 + 2IO_3^- + 4H_2O = 5SO_4^{2-} + I_2 + 8H^+$$

Step 11. The equation is now balanced, with a charge of -2 on both sides, as well as 5S, 2I, 20 O and 8 H atoms on each side.

Example: Balance the following net ionic equation, which occurs in basic solution, using the ion-electron method of balancing.

$$Cr + ClO_4^- = CrO_2^- + ClO_3^-$$

Solution:

Step 1: Assign oxidation numbers.

$$Cr + ClO_4^- = CrO_2^- + ClO_3^-$$
$$0 \quad\ +7 \qquad +3 \qquad +5$$

Cr is oxidized; it increases in oxidation number from 0 to +3.

Cl is reduced; it decreases in oxidation number from +7 to +5.

Step 2: Oxidation $\qquad Cr = CrO_2^-$

Reduction $\qquad ClO_4^- = ClO_3^-$

(Omit steps 3 and 4)

Step 5: Balance oxygen, in basic solution, by adding twice as many OH^- to the oxygen deficient side as H_2O to the other side.

Oxidation $\quad Cr + 4\ OH^- = CrO_2^- + 2H_2O$

Reduction $ClO_4^- + H_2O = ClO_3^- + 2\ OH^-$

(Omit step 6. Hydrogen is already blaanced.)

Step 7: Add electrons to achieve charge balance.

Oxidation $Cr + 4\ OH^- = CrO_2^- + 2H_2O + 3e^-$

Reduction $ClO_4^- + H_2O + 2e^- = ClO_3^- + 2\ OH^-$

Step 8. Multiply the half-reactions by appropriate numbers to make the number of electrons lost equal the number of electrons gained.

Oxidation $2(Cr + 4\ OH^- = CrO_2^- + 2H_2O + 3e^-)$

Reduction $3(ClO_4^- + H_2O + 2e^- = ClO_3^- + 2\ OH^-)$

Step 9. Add the half-reactions, cancelling out the electrons.

Oxidation $2Cr + 8\ OH^- = 2CrO_2^- + 4H_2O + 6e^-$

Reduction $3ClO_4^- + 3H_2O + 6e^- = 3ClO_3^- + 6\ OH^-$

$$2Cr + 3ClO_4^- + 8\ OH^- + 3H_2O + \cancel{6e^-} =$$
$$2CrO_2^- + 3ClO_3^- + 4H_2O + 6\ OH^- + \cancel{6e^-}$$

Step 10. Cancel $3H_2O$ and $6\ OH^-$ from both sides.

$$2Cr + 3ClO_4^- + 2\ OH^- = 2CrO_2^- + 3ClO_3^- + H_2O$$

Step 11. The equation is balanced, with the smallest coefficients possible. There is a charge of -5 on both sides, as well as $2Cr$, $3Cl$, $14\ O$ and $2H$ atoms on each side.

PROBLEM SET

1. Define the terms "oxidation" and "reduction" in terms of:

    a) loss and gain of electrons
    b) increase and decrease in oxidation number

2. Define the terms "oxidizing agent" and "reducing agent" in terms of:

a) loss and gain of electrons
b) increase and decrease in oxidation number
c) substance oxidized and substance reduced

3. Balance each of the following redox equations using the oxidation number method:

a) $HNO_3 + H_2S = H_2SO_4 + NO_2 + H_2O$

b) $FeCl_3 + H_2S = FeCl_2 + HCl + S$

c) $TeO_2 + HBrO_3 + H_2O = H_6TeO_6 + Br_2$

d) $NO + H_5IO_6 = HNO_3 + HIO_3 + H_2O$

4. Balance each of the following redox equations using the ion-electron half-reaction method:

a) $Cl^- + Cr_2O_7^{2-} = Cl_2 + Cr^{3+}$ (in acidic solution)

b) $MnO_4^- + Sn^{2+} = Mn^{2+} + Sn^{4+}$ (in acidic solution)

c) $S + NO_3^- = SO_4^{2-} + NO_2^-$ (in basic solution)

d) $MnO_4^- + H_2C_2O_4 = Mn^{2+} + CO_2$ (in acidic solution)

MULTIPLE CHOICE EXERCISES

1. Which of the following pairings of terms is <u>incorrectly</u> matched?

a) process of oxidation -- loss of electrons
b) substance oxidized -- gain of electrons
c) reducing agent -- loss of electrons
d) substance reduced -- gain of electrons

2. Which of the following statements concerning oxidation and reduction is <u>incorrect</u>?

a) reduction is the gain of electrons by a chemical species
b) an oxidizing agent is a substance that accepts electrons in a redox process
c) a reducing agent causes reduction by losing electrons
d) oxidation is the process that an oxidizing agent undergoes

3. An element's oxidation number:

   a) is the same as its number of valence electrons
   b) equals the number of electrons in its outermost subshell
   c) is calculated using an arbitrary set of rules
   d) is equal to the charge on the ion the element is found in

4. The oxidation number of Cr in $K_2Cr_2O_7$ is:

   a) +1
   b) +3
   c) +6
   d) +12

5. The proper assignment of oxidation numbers to the elements in the ion NH_4^+ is:

   a) 0 for N and +1 for H
   b) −2 for N and +3 for H
   c) −3 for N and +1 for H
   d) +3 for N and −1 for H

6. Select the incorrect pairing of a compound with the oxidation number of the underlined element:

   a) $H\underline{N}O_3$ -- (+5)

   b) $H_2\underline{S}O_4$ -- (+6)

   c) $H\underline{Cl}O_3$ -- (+5)

   d) $Ca\underline{C}O_3$ -- (+3)

7. Which of the following pairing of terms is _incorrect_?

   a) oxidizing agent -- decrease in oxidation number
   b) substance reduced -- increase in oxidation number
   c) reducing agent -- increase in oxidation number
   d) process of reduction -- decrease in oxidation number

8. In a redox reaction the substance oxidized:

   a) contains an element which decreases in oxidation number
   b) is also the oxidizing agent
   c) always loses electrons
   d) must contain oxygen

9. Which substance is the reducing agent in the redox reaction

$$Mg + CuSO_4 \longrightarrow Cu + MgSO_4$$

a) Mg
b) $CuSO_4$
c) Cu
d) $MgSO_4$

10. In the redox reaction $2FeCl_3 + SnCl_2 \longrightarrow 2FeCl_2 + SnCl_4$:

a) $SnCl_2$ is the oxidizing agent
b) $FeCl_3$ is the substance oxidized
c) Sn undergoes an oxidation number change of 2 units
d) Fe undergoes an oxidation number change of 6 units

11. Which of the following reactions is a metathetical reaction:

a) $2FeCl_3 + SnCl_2 \longrightarrow 2FeCl_2 + SnCl_4$

b) $NaCl + AgNO_3 \longrightarrow AgCl + NaNO_3$

c) $2Mg + O_2 \longrightarrow 2MgO$

d) $2KClO_3 \longrightarrow 2KCl + 3 O_2$

12. Which of the following reactions is incorrectly classified:

a) $BaCl_2 + Na_2CO_3 \longrightarrow BaCO_3 + 2NaCl$ (synthesis)

b) $CaCO_3 \longrightarrow CaO + CO_2$ (decomposition)

c) $CO_2 + H_2O \longrightarrow H_2CO_3$ (synthesis)

d) $2NH_3 \longrightarrow 3H_2 + N_2$ (decomposition)

13. Which of the following reactions is a "double replacement" reaction:

a) $2KClO_3 \longrightarrow 2KCl + 3 O_2$

b) $Mg + Ni(NO_3)_2 \longrightarrow Ni + Mg(NO_3)_2$

c) $AgNO_3 + HCl \longrightarrow AgCl + HNO_3$

d) $2FeCl_3 + SnCl_2 \longrightarrow 2FeCl_2 + SnCl_4$

14. In balancing the following equation using the oxidation number method, what will be the values of "a" and "b" on the bottom bracket?

$$\overset{\displaystyle 4(+3)}{\overbrace{\text{Fe} + \underset{\displaystyle a\,[\,2(b)\,]}{\underbrace{O_2 + \text{HCl} \longrightarrow \text{FeCl}_3 + H_2O}}}}$$

a) 3 and +2
b) 2 and −1
c) 3 and −2
d) 2 and −3

15. How many electrons are lost or gained by each molecule of HPO_3 in the reaction:

$$HPO_3 + C \longrightarrow H_2 + CO + P_4$$

a) gains five electrons
b) loses six electrons
c) gains two electrons
d) gains six electrons

16. Which of the following is a correctly balanced oxidation half-reaction?

a) $S + 2e^- \longrightarrow S^{2-}$

b) $Cu^{2+} + 3e^- \longrightarrow Cu^+$

c) $2Cl^- \longrightarrow Cl_2 + 2e^-$

d) $Cr(OH)_3 \longrightarrow CrO_4^{2-} + 4e^-$

17. The balanced full-equation obtained by adding the following two balanced half-reactions is:

$$2I^- \longrightarrow I_2 + 2e^-$$

$$NO_3^- + 4H^+ + 3e^- \longrightarrow NO + 2H_2O$$

a) $2I^- + NO_3^- + 4H^+ \longrightarrow I_2 + NO + 2H_2O$

b) $4I + NO_3^- + 4H^+ \longrightarrow 2I_2 + NO + 2H_2O$

c) $6I + 2NO_3^- + 8H^+ \longrightarrow 3I_2 + 2NO + 4H_2O$

d) $2I + 2NO_3^- + 8H^+ \longrightarrow I_2 + 2NO + 4H_2O$

18. When the half-reaction $BrO_4^- \longrightarrow Br^-$ (basic solution) is correct balanced:

    a) the OH^- and H_2O are both on the left side of the equation
    b) the OH^- and H_2O are both on the right side of the equation
    c) the OH^- is on the left and the H_2O on the right side of the equation
    d) the OH^- is on the right and the H_2O on the left side of the equation

ANSWERS TO PROBLEMS SET

1. Oxidation is: (a) the loss of electrons or (b) the increase in oxidation number.

   Reduction is: (a) the gain of electrons or (b) the decrease in oxidation number.

2. An oxidizing agent: a) gains electrons, b) decreases in oxidation number, c) is the substance reduced.

   A reducing agent: a) loses electrons, b) increases in oxidation number, c) is the substance oxidized.

3. a) **Steps 1, 2, (3)**

$$\overset{-2\ (+8)\ +6}{\overbrace{}}$$

$$\underset{+1+5-2 \quad +1\ -2 \quad +1+6-2 \quad +4-2 \quad +1-2}{HNO_3 + H_2S = H_2SO_4 + NO_2 + H_2O}$$

$$\underset{+5 \qquad (-1) \qquad +4}{\underbrace{}}$$

Steps 4, 5

$$\overset{(+8)}{\overbrace{}}$$

$$8HNO_3 + H_2S = H_2SO_4 + 8NO_2 + H_2O$$

$$\underset{8(-1)}{\underbrace{}}$$

Steps 6, on

$$8HNO_3 + H_2S = H_2SO_4 + 8NO_2 + 4H_2O$$

b) Steps 1, 2, (3)

$$\overset{+3}{\underset{+3-1}{FeCl_3}} + \overset{(-1)}{\underset{+1-2}{H_2S}} = \overset{+2}{\underset{+2-1}{FeCl_2}} + \underset{+1-1}{HCl} + \underset{0}{S}$$

$$-2 \qquad\qquad (+2) \qquad\qquad 0$$

Steps 4, 5

$$2FeCl_3 + \overset{2(-1)}{H_2S} = 2FeCl_2 + HCl + S$$

$$(+2)$$

Steps 6, on

$$2FeCl_3 + H_2S = 2FeCl_2 + 2HCl + S$$

c) Steps 1, 2

$$\overset{+4}{\underset{+4-2}{TeO_2}} + \overset{(+2)}{\underset{+1+5-2}{HBrO_3}} + \underset{+1-2}{H_2O} = \overset{+6}{\underset{+1+6-2}{H_6TeO_6}} + \underset{0}{Br_2}$$

$$+5 \qquad\qquad (-5) \qquad\qquad 0$$

Step 3

$$TeO_2 + \overset{(+2)}{HBrO_3} + H_2O = H_6TeO_6 + Br_2$$

$$2(-5)$$

Steps 4, 5

$$5TeO_2 + 2\overset{5(+2)}{HBrO_3} + H_2O = 5H_6TeO_6 + Br_2$$

$$2(-5)$$

Steps 6, on

$$5TeO_2 + 2HBrO_3 + 14H_2O = 5H_6TeO_6 + Br_2$$

d) **Steps 1, 2, (3)**

$$\overset{+2}{NO} + \overset{(+3)}{H_5IO_6} = \overset{+5}{HNO_3} + HIO_3 + H_2O$$

$$\begin{array}{ccccc} +2-2 & +1+7-2 & +1+5-2 & +1+5-2 & +1-2 \end{array}$$

$$\begin{array}{ccc} +7 & (-2) & +5 \end{array}$$

Steps 4, 5

$$2NO + 3H_5IO_6 \overset{2(+3)}{=} 2HNO_3 + 3HIO_3 + H_2O$$

$$3(-2)$$

Step 6, on

$$2NO + 3H_5IO_6 = 2HNO_3 + 3HIO_3 + 5H_2O$$

4. a) **Step 1.** Determine oxidation numbers.

$$\overset{-}{Cl} + Cr_2O_7^{2-} = Cl_2 + Cr^{3+} \text{ (acidic solution)}$$

$$\begin{array}{cccc} -1 & +6 \ -2 & 0 & +3 \end{array}$$

Steps 2, 3. Write half-reactions for substances oxidized and reduced. Balance each half-reaction for number of atoms of the elements oxidized and reduced.

Oxidation $\qquad 2Cl^- = Cl_2$

Reduction $\qquad Cr_2O_7^{2-} = 2Cr^{3+}$

Steps (4), 5. Balance for oxygen.

Oxidation $\qquad 2Cl^- = Cl_2$

Reduction $\qquad Cr_2O_7^{2-} = 2Cr^{3+} + 7H_2O$

Step 6. Balance for hydrogen.

Oxidation $\qquad 2Cl^- = Cl_2$

Reduction $\quad Cr_2O_7^{2-} + 14H^+ = 2Cr^{3+} + 7H_2O$

Step 7. Balance for charge.

Oxidation $\quad 2Cl^- = Cl_2 + 2e^-$

Reduction $\quad Cr_2O_7^{2-} + 14H^+ + 6e^- = 2Cr^{3+} + 7H_2O$

Step 8. Multiply the half-reactions by the numbers necessary to make the number of electrons lost equal the number of electrons gained.

Oxidation $\quad 3(2Cl^- = Cl_2 + 2e^-)$

Reduction $\quad Cr_2O_7^{2-} + 14H^+ + 6e^- = 2Cr^{3+} + 7H_2O$

Step 9, on. Add the two half-reactions.

Oxidation $\quad 6Cl^- = 3Cl_2 + 6e^-$

Reduction $\quad Cr_2O_7^{2-} + 14H^+ + 6e^- = 2Cr^{3+} + 7H_2O$

$$Cr_2O_7^{2-} + 6Cl^- + 14H^+ + \cancel{6e^-} = 2Cr^{3+} + 3Cl_2 + 7H_2O + \cancel{6e^-}$$

The equation is balanced. Both sides show a charge of +6 and 2Cr, 6Cl, 14H and 7 O atoms.

$$Cr_2O_7^{2-} + 6Cl^- + 14H^+ = 2Cr^{3+} + 3Cl_2 + 7H_2O$$

b) **Step 1.** $\quad MnO_4^- + Sn^{2+} = Mn^{2+} + Sn^{4+}$ (acid solution)

$\qquad\qquad +7-2 \qquad +2 \qquad\qquad +2 \qquad +4$

Steps 2, (3, 4)

Oxidation $\quad Sn^{2+} = Sn^{4+}$

Reduction $\quad MnO_4^- = Mn^{2+}$

Step 5. Balance for oxygen.

Oxidation $\quad Sn^{2+} = Sn^{4+}$

Reduction $\quad MnO_4^- = Mn^{2+} + 4H_2O$

Step 6. Balance for hydrogen.

Oxidation $\quad Sn^{2+} = Sn^{4+}$

Reduction $\quad MnO_4^- + 8H^+ = Mn^{2+} + 4H_2O$

Step 7. Balance for charge.

Oxidation $Sn^{2+} = Sn^{4+} + 2e^-$

Reduction $MnO_4^- + 8H^+ + 5e^- = Mn^{2+} + 4H_2O$

Step 8. Make the number of electrons lost equal the number of electrons gained.

Oxidation $5(Sn^{2+} = Sn^{4+} + 2e^-)$

Reduction $2(MnO_4^- + 8H^+ + 5e^- = Mn^{2+} + 4H_2O$

Step 9 (on). Add the half-reactions.

Oxidation $5Sn^{2+} = 5Sn^{4+} + 10e^-$

Reduction $2MnO_4^- + 16H^+ + 10e^- = 2Mn^{2+} + 8H_2O$

$$2MnO_4^- + 16H^+ + 5Sn^{2+} + \cancel{10e^-} = 2Mn^{2+} + 8H_2O + 5Sn^{4+} + \cancel{10e^-}$$

The equation is balanced, each side shows a charge of +24, and 2Mn, 5Sn, 16H and 8 O atoms.

$$2MnO_4^- + 16H^+ + 5Sn^{2+} = 2Mn^{2+} + 8H_2O + 5Sn^{4+}$$

c) **Step 1.** $S + NO_3^- = SO_4^{2-} + NO_2^-$ (basic solution)
       0   +5   +6   +3

Steps 2, (3, 4).

Oxidation $S = SO_4^{2-}$

Reduction $NO_3^- = NO_2^-$

Step 5, (6). Balance the oxygen by adding OH^- and H_2O where needed.

Oxidation $S + 8 OH^- = SO_4^{2-} + 4H_2O$

Reduction $NO_3^- + H_2O = NO_2^- + 2 OH^-$

Step 7. Balance for charge.

Oxidation $S + 8 OH^- = SO_4^{2-} + 4H_2O + 6e^-$

Reduction $NO_3^- + H_2O + 2e^- = NO_2^- + 2 OH^-$

Step 8. Make the electrons lost equal the electrons gained.

Oxidation $\quad S + 8\ OH^- = SO_4^{2-} + 4H_2O + 6e^-$

Reduction $\quad 3(NO_3^- + H_2O + 2e^- = NO_2^- + 2\ OH^-)$

Step 9. Add the half-reactions.

Oxidation $\quad S + 8\ OH^- = SO_4^{2-} + 4H_2O + 6e^-$

Reduction $\quad 3NO_3^- + 3H_2O + 6e^- = 3NO_2^- + 6\ OH^-$

$$S + 3NO_3^- + 8\ OH^- + 3H_2O + \cancel{6e^-} =$$

$$SO_4^{2-} + 3NO_2^- + 4H_2O + 6\ OH^- + \cancel{6e^-}$$

Step 10. Cancel substances found on both sides; in this case 6 OH^- and $3H_2O$.

$$S + 3NO_3^- + 2\ OH^- = SO_4^{2-} + 3NO_2^- + H_2O$$

The equation is balanced, showing (on both sides) a charge of −5, and 1 S, 3 N, 11 O and 2 H atoms.

d) **Step 1.** $\quad MnO_4^- + H_2C_2O_4 = Mn^{2+} + CO_2$ (acidic
$+7 +3 +2 \phantom{Mn^{2+}} +4 $ solution)

Step 2. Oxidation $\quad H_2C_2O_4 = CO_2$

 Reduction $\quad MnO_4^- = Mn^{2+}$

Step 3, (4) $\quad$ Oxidation $\quad H_2C_2O_4 = 2CO_2$

 Reduction $\quad MnO_4^- = Mn^{2+}$

Step 5. Balance oxygen.

Oxidation $\quad H_2C_2O_4 = 2CO_2$

Reduction $\quad MnO_4^- = Mn^{2+} + 4H_2O$

Step 6. Balance hydrogen.

Oxidation $\quad H_2C_2O_4 = 2CO_2 + 2H^+$

Reduction $\quad MnO_4^- + 8H^+ = Mn^{2+} + 4H_2O$

Step 7. Balance charge.

Oxidation $\quad H_2C_2O_4 \;=\; 2CO_2 + 2H^+ + 2e^-$

Reduction $\quad MnO_4^- + 8H^+ + 5e^- \;=\; Mn^{2+} + 4H_2O$

Step 8. Make the electrons lost equal the electrons gained.

Oxidation $\quad 5(H_2C_2O_4 \;=\; 2CO_2 + 2H^+ + 2e^-)$

Reduction $\quad 2(MnO_4^- + 8H^+ + 5e^- \;=\; Mn^{2+} + 4H_2O)$

Step 9. Add half-reactions.

Oxidation $\quad 5H_2C_2O_4 \;=\; 10\ CO_2 + 10\ H^+ + 10e^-$

Reduction $\quad 2MnO_4^- + 16H^+ + 10e^- \;=\; 2Mn^{2+} + 8H_2O$

$$5H_2C_2O_4 + 2MnO_4^- + 16H^+ + \cancel{10e^-} =$$
$$10\ CO_2 + 2Mn^{2+} + 10\ H^+ + 8H_2O + \cancel{10e^-}$$

Step 10. Cancel 10 H^+ on each side.

$$5H_2C_2O_4 + 2MnO_4^- + 6H^+ \;=\; 10\ CO_2 + 2Mn^{2+} + 8H_2O$$

The equation is now balanced, both sides showing a charge of +4 and 10 C, 2 Mn, 16 H and 28 O atoms.

ANSWERS TO MULTIPLE CHOICE EXERCISES

1.	b	10.	c
2.	d	11.	b
3.	c	12.	a
4.	c	13.	c
5.	c	14.	a
6.	d	15.	a
7.	b	16.	c
8.	c	17.	c
9.	a	18.	d

CHAPTER 16
Reaction Rates and Chemical Equilibrium

REVIEW OF CHAPTER OBJECTIVES

1. <u>Explain what is meant by a reaction rate</u> (section 16.1).

   Reaction rate is the rate or speed at which reactants are consumed or products are produced during a chemical reaction.

   An example of a slow reaction is the rusting of iron. In this reaction the reactants are consumed over a long period of time.

   An example of a fast reaction is the burning of gun powder. In this reaction, the reactants are consumed very rapidly over a very short time period.

2. <u>Describe how reactions occur using collision theory</u> (section 16.1).

   Collision theory is an explanation of how a chemical reaction occurs.

   Collision theory contains three fundamental postulates:

   1. <u>Reactant particles must collide with each other in order for a reaction to occur</u>.

      When reactions involve two or more reactants, collision theory assumes that the reactant molecules, ions or atoms must come into contact (collide) with each other for reaction to occur. The validity of this assumption is fairly obvious. Reactants cannot react with each other if they are miles apart.

   2. <u>Colliding particles must collide with a certain minimum total amount of energy if the collision is to result in a reaction</u>.

This postulate indicates that for reaction to occur, the sum of the kinetic energy of the colliding particles must add up to a certain minimum value. This minimum combined kinetic energy the reactant particles must possess for their collision to result in a reaction is called the activation energy. Every chemical reaction has a different activation energy.

Not all collisions result in product formation. If the activation energy is exceeded, product is formed. If the activation energy is not exceeded, then the reactant particles rebound from the collision and remain unchanged.

3. In some cases, reactants must be oriented in a specific way upon collision if a reaction is to occur.

Even when activation energy requirements are met, some collisions between reactant particles still do not result in product formation. For nonspherical molecules and polyatomic ions, their orientation relative to each other at the moment of collision is a factor in determining whether a collision is effective.

3. Draw potential energy diagrams for exothermic and endothermic reactions (section 16.2).

The relationship between the activation energy for a chemical reaction and the total potential energy of the reactants and products can be illustrated graphically using potential energy diagrams.

Exothermic reaction -- the average total energy of the reactants is higher than the average total energy of the products, and, thus, excess energy is released to the surroundings.

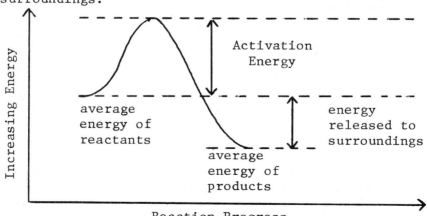

In this diagram, the total average energy of the products is lower than the total average energy of the reactants. This difference in energy is released to the surroundings and the reaction is said to be exothermic.

Endothermic reaction -- the average total energy of the reactants is less than the average total energy of the products, and the energy necessary for the reaction has been absorbed from the surroundings.

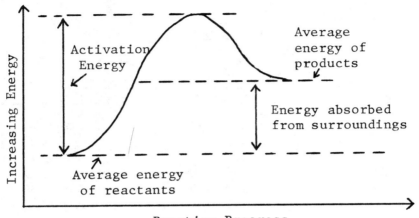

In this diagram, the total average energy of the reactants is lower than the total average energy of the products. This difference in energy is absorbed from the surroundings and the reaction is said to be endothermic.

4. List the four factors affecting the rate of reaction and explain how they operate in terms of collision theory (section 16.3).

Four factors that affect the rate of reaction are:

1. Physical nature of the reactants.

   Physical nature refers to the physical state (solid, liquid or gas) and the size of the particles for solids.

   If the reactants are all in the same physical state, the order of decreasing reactivity is:

   1) Gases

      There is the greatest freedom of movement of particles in the gaseous state. As a result, there is a greater frequency of collision between reactant particles.

2) Liquids

There is more freedom of movement of particles in the liquid state than in the solid state, and less freedom of movement in the liquid state than in the gaseous state. As a result, the rate of reaction in the liquid state is between that of solids and gases.

3) Solids

There is the least freedom of movement in the solid state. As a result, the rate of reaction of solids is the slowest of the three physical states.

In reactions where solids are involved, reaction occurs at the boundary surface between the reactants. The greater the amount of surface area, the faster the reaction rate. Subdividing a solid into smaller particles increases the surface area and the reaction rate by increasing the number of productive collisions.

2. Concentration of reactants

An increase in the concentration of a reactant causes an increase in the rate of reaction. Increasing the concentration of a reactant means that there are more particles of that reactant present in the reaction mixture and, therefore, there is a greater chance for collisions between this reactant and other reactant particles.

3. Temperature

In a reaction taking place at a high temperature the greater average kinetic energy of the molecules results in increased speed for the molecules. This increased speed of the molecules causes more collisions to take place in a given time. Also, since the energy of the colliding molecules is greater, a larger fraction of the collisions will result in reaction from the point of view of activation energy.

As a rule of thumb, it has been found that rate of reaction doubles for every $10^{\circ}C$ increase in temperature.

4. Presence of a catalyst

A catalyst is a substance that, when added to a reaction mixture, increases the rate of the reaction but which, itself, remains unchanged after the reaction is completed.

Catalysts increase reaction rates by providing alternate reaction pathways with lower activation energies than the original uncatalyzed pathway.

5. Explain what is meant by chemical equilibrium and the conditions necessary for the attainment of such a state (section 16.4).

Chemical equilibrium is the process where two opposing chemical reactions occur simultaneously at the same rate.

Equilibrium involves a balance between opposing processes. Two previously described examples of physical equilibrium are vaporization and saturation.

Chemical equilibrium always involves a reversible reaction:

$$\text{reactants} \underset{\text{reverse}}{\overset{\text{forward}}{\rightleftarrows}} \text{products}$$

A reversible reaction is one in which the reactants are reacting to form products, and the products formed are reacting to reform the reactants. Reversible reactions are indicated by using double arrows when writing the chemical equation.

Chemical equilibrium is reached when the rate of the reverse reaction is equal to the rate of the forward reaction. At that point the concentrations of reactants and products are constant.

An important point about chemical equilibrium is that the rates of the forward and reverse reactions are equal, but the concentrations of reactants and products are not equal. Equilibrium may exist with a small amount of product present or a large amount of product present, or any amount in between.

6. Write the equilibrium constant expression for a reaction given a balanced chemical equation (section 16.5).

An equilibrium constant is the product of the molar concentrations of the products, each raised to the power of its respective coefficient in the balanced chemical equation, divided by the product of the molar concentrations of the reactants, each raised to the power of its respective coefficient in the balanced chemical equation.

The following conventions are followed in writing equilibrium constant expressions:

1. Concentrations of gases and aqueous solutions are always expressed in moles per liter, indicated by using square brackets [ ].

2. Product concentrations are always placed in the numerator of the equilibrium constant expression.

3. Reactants are always placed in the denominator of the equilibrium constant expressions.

$$K_{eq} = \frac{products}{reactants}$$

4. The powers to which the concentrations are raised are always determined by the coefficients in the balanced chemical equation for the reaction.

5. Pure solids and liquids are omitted from equilibrium constant expressions because they have fixed concentrations and are included in the value of K_{eq}.

To illustrate the writing of an equilibrium constant expression, consider a general gas phase reaction in which a moles of A and b moles of B (the reactants) react to produce c moles of C and d moles of D (the products).

Written in equation form, we have:

$$aA_{(g)} + bB_{(g)} = cC_{(g)} + dD_{(g)}$$

An equal sign (=) is sometimes used in place of the double arrow.

The equilibrium constant expression for this reaction is:

$$K_{eq} = \frac{products}{reactants} = \frac{[C]^c[D]^d}{[A]^a[B]^b}$$

Reviewing the conventions used in writing this

expression, we see that:

1. The concentrations $[A]$, $[B]$, $[C]$ and $[D]$ are enclosed in brackets.

2. The product concentrations, $[C]$ and $[D]$ are written in the numerator.

3. The reactant concentrations, $[A]$ and $[B]$ are written in the denominator.

4. The powers a, b, c and d to which the concentrations are raised $[A]^a$, $[B]^b$, $[C]^c$ and $[D]^d$ are the coefficients a, b, c and d in the balanced chemical equation.

5. There are no solids or liquids so all reactant and product concentrations appear in the equilibrium constant expression.

Example: Write the equilibrium constant expression for each of the following reactions:

a) $3H_{2(g)} + N_{2(g)} \rightleftharpoons 2NH_{3(g)}$

b) $C_2H_{4(g)} + 3 O_{2(g)} \rightleftharpoons 2CO_{2(g)} + 2H_2O_{(g)}$

c) $Ag^+_{(aq)} + Cl^-_{(aq)} \rightleftharpoons AgCl_{(s)}$

d) $Al_2(SO_4)_{3(s)} \rightleftharpoons 2Al^{3+}_{(aq)} + 3SO_4^-{}_{(aq)}$

Solution:

a) All of the substances involved in this reaction are gases. Therefore, each reactant and product will appear in the equilibrium constant expression.

$$K_{eq} = \frac{[NH_3]^2}{[H_2]^3[N_2]}$$

b) The product water is a gas and is included in the equilibrium constant expression. The concentration of water vapor varies, while that of water as a liquid does not vary and is never included in equilibrium constant expressions.

$$K_{eq} = \frac{[CO_2]^2[H_2O]^3}{[C_2H_4][O_2]^3}$$

338   Reaction Rates and Chemical Equilibrium

c) The product AgCl is a solid and, therefore, will not appear in the equilibrium constant expression. Since AgCl is the only product, no terms appear in the numerator and a 1 is used to indicate that the $[Ag^+]$ and $[Cl^-]$ are in the denominator.

$$K_{eq} = \frac{1}{[Ag^+][Cl^-]}$$

d) The equilibrium constant expression will contain the concentrations of the ions in aqueous solution but not the concentration of the solid from which they are produced. This means there is no denominator in the equilibrium constant expression.

$$K_{eq} = [Al^{3+}]^2 [SO_4^{2-}]^3$$

7. <u>Qualitatively interpret values of K_{eq} in terms of position of the equilibrium</u> (section 16.5).

The magnitude of the equilibrium constant for a reaction gives information about the quantity of reactants consumed when equilibrium is reached.

To understand the meaning of equilibrium constants it is important to realize that equilibrium reactions do not go to completion. That is, all reactants are not completely converted to products. At equilibrium, both reactant and product species are present. For example, consider the equilibrium:

$$H_2 + I_2 \rightleftharpoons 2HI$$

Equal amounts of H_2 and I_2 are sealed in a container and allowed to react until equilibrium is reached. If we then analyze the equilibrium mixture we will find all three substances (H_2, I_2 and HI) present.

If the reaction went to completion, we would only find product (HI) present.

If no reaction takes place, we would find only H_2 and I_2 present.

The point of equilibrium for this reaction lies somewhere between all (HI) and all (H_2 and I_2).

The magnitude of K_{eq} tells us where the point of equilibrium lies.

$$K_{eq} = \frac{products}{reactants} = \frac{[HI]^2}{[H_2][I_2]}$$

If K_{eq} has a large value (more than 10^3), there is more product (HI) than reactants (H_2 and I_2). This means the numerical value of the numerator is greater than the denominator in the equilibrium constant expression, and the equilibrium point is on the product side of the reaction.

If K_{eq} has a small value (less than 10^{-3}), there are more reactants (H_2 and I_2) than product (HI). This means the numerical value of the numerator is smaller than the denominator in the equilibrium constant expression, and the equilibrium point is on the reactant side of the reaction.

For equilibrium conditions where K_{eq} has a value close to one, appreciable concentrations of both products and reactants are present at the equilibrium point.

Table 16-1 on page 442 of the textbook summarizes the relationship between equilibrium constant magnitude, concentrations of products and reactants at equilibrium and the equilibrium position.

Example: Describe qualitatively the position of the equilibrium for each of the following reactions:

a) $2SO_{2(g)} + O_{2(g)} \rightleftharpoons 2SO_{3(g)}$ $K_{eq} = 1.0 \times 10^{150}$

b) $CO_{(g)} + 2H_{2(g)} \rightleftharpoons CH_3OH$ $K_{eq} = 3.0 \times 10^5$

c) $C_{(s)} + H_2O_{(g)} \rightleftharpoons H_{2(g)} + CO_{(g)}$ $K_{eq} = 5 \times 10^{-8}$

d) $4HCl_{(g)} + O_{2(g)} \rightleftharpoons 2H_2O_{(g)} + 2Cl_{2(g)}$ $K_{eq} = 2 \times 10^{-25}$

Solution:

    a) The position of equilibrium is far to the right. For all intents and purposes, this reaction has gone to completion. With such a large K_{eq}, only traces of reactants would be present in the equilibrium mixture.

    b) The equilibrium lies to the right, that is, more products are present than reactants at equilibrium.

    c) The reaction lies to the left, that is, more

reactants are present than products. The
reaction has occurred only to a slight extent.

    d) The position of equilibrium lies far to the left.
Essentially, the reaction has not occurred. Only
a minute amount of product is in the equilibrium
mixture.

8. <u>State Le Chatelier's Principle</u> (section 16.6).

Le Chatelier's Principle states that when a reaction, in
a state of equilibrium, is subjected to a stress, such as
changes in pressure, temperature or concentration, the
tendency is to shift the equilibrium in such a way as to
relieve the stress, and a new equilibrium will be
established under the new conditions.

9. <u>Use Le Chatelier's Principle to predict the effect that
concentration, temperature and pressure changes will have
on an equilibrium system</u> (section 16.6).

A change in concentration of one substance in an
equilibrium mixture affects the equilibrium as follows:

1. Adding more reactant to the equilibrium will cause
the reaction to shift to the right and result in an
increase in the amount of products present in the new
equilibrium.

2. Adding more product to the equilibrium mixture will
cause a shift to the left and result in an increase
in the amount of reactants present in the new
equilibrium.

3. Reactants or products may also be removed from the
equilibrium mixture, and the reaction will shift in
the direction that will produce more of the substance
that was removed.

Equilibrium mixtures containing one or more gases are
affected by a change in pressure as follows:

1. A pressure change affects the equilibrium only if the
total number of moles of gaseous reactants is
different than the total number of moles of gaseous
products in the balanced chemical equation.
According to Le Chatelier's Principle, the stress of
increased pressure is relieved by decreasing the
number of moles of gaseous substances in the
equilibrium mixture. This is accomplished by the
reaction shifting to the side of the equation which

contains the fewer number of moles of gaseous
substance.

2. A decrease in pressure is relieved by increasing the
    number of moles of gaseous substances in the
    equilibrium mixture. This is accomplished by the
    reaction shifting to the side of the equation which
    contains the larger number of moles of gaseous
    substance.

Changes in the temperature of an equilibrium system
affect the equilibrium in the following manner:

1. Raising the temperature favors the reaction that
    absorbs heat (an endothermic reaction). Product
    formation is favored by temperature increase if the
    forward reaction is endothermic.

2. Lowering the temperature favors the reaction that
    gives off heat (an exothermic reaction). Product
    formation is favored by lowering the temperature if
    the forward reaction is exothermic.

A catalyst has no effect on a system at equilibrium, but
shortens the time it takes to reach equilibrium.

Example: How will the gaseous phase equilibrium

$$N_{2(g)} + 3H_{2(g)} \rightleftharpoons 2NH_{3(g)} + heat$$

be affected by:

a) adding NH_3

b) removing H_2

c) increasing the temperature

d) decreasing the pressure (increasing the
   volume)

Solution:

a) The equilibrium will shift to the left in an
   attempt to use up the extra NH_3 which has been
   placed in the system.

b) The equilibrium will shift to the left in an
   attempt to replace the H_2 which was removed.

c) This is an exothermic reaction so an increase in
   temperature will cause a shift to the left.

d) The system shifts to the left in an attempt to produce more moles of gaseous reactant which will increase the pressure. In going to the left, the reaction produces 4 moles of gaseous reactant for every 2 moles of gaseous product consumed.

10. <u>State several ways in which a reaction may be "forced to completion"</u> (section 16.7).

Reactions which would ordinarily reach a state of equilibrium can be forced to completion by using experimental conditions which place a "continual stress" on the potential equilibrium.

A reaction can be "forced to completion" in the following ways:

1. Continuous removal of one or more products.

2. Use of a large excess of one of the reactants.

3. Increase of pressure if there is a smaller number of moles of product than reactant.

4. Decrease in temperature if the reaction is an exothermic reaction.

5. Increase in temperature if the reaction is an endothermic reaction.

PROBLEM SET

1. What are the three basic postulates of collision theory?

2. What are the four factors which influence the rate of a chemical reaction?

3. Write the equilibrium constant expression for each of the following reactions?

a) $CO_{(g)} + H_2O_{(g)} \rightleftharpoons CO_{2(g)} + H_{2(g)}$

b) $2SO_{2(g)} + O_{2(g)} \rightleftharpoons 2SO_{3(g)}$

c) $N_2O_{4(g)} \rightleftharpoons 2NO_{2(g)}$

d) $2H_2O_{(g)} \rightleftharpoons 2H_{2(g)} + O_{2(g)}$

e) $2 O_{3(g)} \rightleftharpoons 3 O_{2(g)}$

f) $CO_{(g)} + 2H_{2(g)} \rightleftharpoons CH_3OH_{(g)}$

g) $CaCO_{3(s)} \rightleftharpoons CaO_{(s)} + CO_{2(g)}$

h) $C_{(s)} + H_2O_{(g)} \rightleftharpoons H_{2(g)} + CO_{(g)}$

4. Describe qualitatively the position of the equilibrium for each of the following reactions:

a) $H_{2(g)} + Cl_{2(g)} \rightleftharpoons 2HCl \qquad K_{eq} = 2.0 \times 10^{80}$

b) $S_{(s)} + O_{2(g)} \rightleftharpoons SO_2 \qquad K_{eq} = 5.0 \times 10^{6}$

c) $CuO_{(s)} + H_{2(g)} \rightleftharpoons Cu_{(s)} + H_2O_{(g)} \qquad K_{eq} = 1.0 \times 10^{-5}$

d) $SnO_{2(s)} + 2CO_{(g)} \rightleftharpoons Sn_{(s)} + 2CO_{2(g)} \qquad K_{eq} = 3.0 \times 10^{-30}$

5. List the three changes in conditions that can affect an equilibrium, according to Le Chatelier's Principle.

6. For the reaction

$$4NH_{3(g)} + 7 O_{2(g)} \rightleftharpoons 4NO_{2(g)} + 6H_2O_{(g)} + heat$$

determine the direction that the equilibrium will be shifted by the following changes:

a) Adding NH_3 e) Increasing the temperature

b) Removing O_2 f) Decreasing the temperature

c) Adding NO_2 g) Increasing the pressure

d) Removing H_2O h) Decreasing the pressure

7. For the reaction

$$2SO_{2(g)} + O_{2(g)} \rightleftharpoons 2SO_{3(g)} + heat$$

determine the direction that the equilibrium will be shifted by the following changes:

a) Increasing the SO_2 concentration

b) Decreasing the O_2 concentration

c) Increasing the SO_3 concentration

d) Increasing the temperature

e) Decreasing the pressure

MULTIPLE CHOICE EXERCISES

1. In most cases before chemical substances can react with one another, their molecules must:

    a) decompose
    b) dissociate
    c) be in the vapor state
    d) collide with one another

2. For a collision between molecules to result in reaction, the molecules must:

    a) possess the same energy
    b) "stick together" for at least 10 seconds
    c) exchange electrons
    d) possess at least a certain minimum energy

3. Which of the following is not true concerning the requirements for a reaction to occur?

    a) molecules must collide with sufficient energy
    b) molecules must be properly oriented at the time of collision
    c) molecules collectively must have kinetic energies whose sum is equal to the activation energy for the reaction
    d) molecules must contain the same number of atoms

4. Activation energy:

    a) is high for reactions which take place rapidly
    b) is an energy valley between reactants and products
    c) is the minimum energy necessary for colliding molecules to react
    d) is the energy given off when reactants collide

5. Which of the following reactions would be predicted to occur most rapidly?

    a) reaction between two solids
    b) reaction between two soluble ionic compounds in aqueous solution
    c) reaction between a solid compound and a liquid compound
    d) reaction between two gases

6. The most important effect of increasing the temperature of a reaction is to:

   a) lower the activation energy
   b) raise the activation energy
   c) increase the number of molecules with the necessary energy to react
   d) decrease the number of molecules with the necessary energy to react

7. What effect does the addition of heat usually have on the rate of a chemical reaction?

   a) increases it
   b) decreases it
   c) may either increase or decrease it depending on reaction conditions
   d) has little or no effect on it

8. Which of the following is not a correct statement about a catalyst?

   a) it changes the reaction pathway
   b) it lowers the energy of activation
   c) it increases the activation energy and increases the rate of the reaction
   d) it is not consumed in the reaction

9. The missing words in the statement "A potential energy diagram shows graphically the relationship between the _____ energy for a chemical reaction and the total _____ energy of the reactants and products" are:

   a) potential and kinetic
   b) potential and activation
   c) activation and kinetic
   d) activation and potential

10. Which of the following is not a correct statement about chemical equilibrium?

   a) reactant concentrations remain constant
   b) product concentrations remain constant
   c) rate of the forward reaction has dropped to zero
   d) reactants are being consumed at the same rate they are being produced

11. Chemical equilibrium can be defined as the point at which:

a) all molecules stop reacting
b) all reactant molecules are used up
c) the product molecules react together at the same rate as reactant molecules
d) product molecules no longer react with each other

12. At equilibrium how do the rates for the forward and reverse reaction compare?

a) the rates are always equal
b) the forward reaction rate is always greater
c) the reverse reaction rate is always greater
d) the forward reaction rate is always double the reverse reaction rate

13. Which of the following is the correct equilibrium expression for the reaction

$$N_{2(g)} + 3H_{2(g)} \rightleftharpoons 2NH_{3(g)} + heat \ ?$$

a) $K_{eq} = \dfrac{\left[NH_3\right]^2}{\left[N_2\right]\left[H_2\right]^3}$

b) $K_{eq} = \left[N_2\right]\left[H_2\right]^3$

c) $K_{eq} = \left[NH_3\right]^2$

d) $K_{eq} = \dfrac{\left[N_2\right]\left[H_2\right]^3}{\left[NH_3\right]^2}$

14. If the value of the equilibrium constant is very large, it means:

a) the equilibrium position is far to the right
b) the equilibrium position is to the right
c) the equilibrium position is to the left
d) the equilibrium position is far to the left

15. When the position of an equilibrium is described as being "far to the right" it means that:

a) very few reactant molecules are present
b) very few product molecules are present
c) the product molecules weigh more than the reactant molecules

d) the product molecules are "very conservative" in their thoughts and actions

16. CO_2 and H_2 are allowed to react until an equilibrium is established as follows:

$$CO_2 + H_2 \rightleftharpoons H_2O + CO$$

What will be the effect on the equilibrium of adding CO_2 to the mixture?

a) the equilibrium will shift to the left
b) H_2 will increase
c) CO and H_2 will increase
d) H_2 will decrease and H_2O will increase

17. Which of the following changes will, according to Le Chatelier's Principle, shift to the left the equilibrium of the reaction

$$N_2 + 3H_2 \rightleftharpoons 2NH_3 + heat$$

a) increase the concentration of N_2
b) decrease the concentration of NH_3
c) increase the temperature
d) increase the concentration of H_2

18. A chemical system is at equilibrium and a catalyst is added. What happens?

a) the position of the equilibrium shifts to the right
b) the position of the equilibrium shifts to the left
c) the position of the equilibrium shifts, but the direction cannot be predicted without knowing what the catalyst is
d) nothing happens

19. Which of the following changes would, according to Le Chatelier's Principle, shift to the right the equilibrium of the reaction

$$heat + 2HI \rightleftharpoons H_2 + I_2$$

a) add a catalyst
b) add H_2
c) remove I_2
d) remove heat

20. Which of the following would force to completion the reaction

$$N_{2(g)} + 3H_{2(g)} \longrightarrow 2NH_{3(g)} + heat$$

a) decrease the pressure
b) increase the temperature
c) decrease the temperature
d) use less H_2

ANSWERS TO PROBLEM SET

1. a) Reactant particles must collide with each other for a reaction to occur.
    b) Colliding particles must collide with a certain minimum amount of energy.
    c) Reactants must be correctly oriented.

2. a) Physical nature of reactants
    b) Concentration of reactants
    c) Temperature
    d) Presence of a catalyst

3. a) $K_{eq} = \dfrac{[CO_2][H_2]}{[CO][O_2]}$

    b) $K_{eq} = \dfrac{[SO_3]^2}{[SO_2]^2[O_2]}$

    c) $K_{eq} = \dfrac{[NO_2]^2}{[N_2O_4]}$

    d) $K_{eq} = \dfrac{[H_2]^2[O_2]}{[H_2O]^2}$

    e) $K_{eq} = \dfrac{[O_2]^3}{[O_3]^2}$

    f) $K_{eq} = \dfrac{[CH_3OH]}{[CO][H_2]^2}$

    g) $K_{eq} = [CO_2]$

    h) $K_{eq} = \dfrac{[H_2][CO]}{[H_2O]}$

4. a) Far to the right
    b) To the right
    c) To the left
    d) Far to the left

5. a) Temperature
    b) Concentration
    c) Pressure

6.   a) Right   e) Left
       b) Left   f) Right
       c) Left   g) Right
       d) Right   h) Left

7.   a) Right
       b) Left
       c) Left
       d) Left
       e) Left

ANSWERS TO MULTIPLE CHOICE EXERCISES

1.	d	11.	c
2.	d	12.	a
3.	d	13.	a
4.	c	14.	a
5.	d	15.	a
6.	c	16.	d
7.	c	17.	c
8.	c	18.	d
9.	d	19.	c
10.	c	20.	c

CHAPTER 17
Nuclear Chemistry

REVIEW OF CHAPTER OBJECTIVES

1. <u>Define the term radioactive</u> (section 17.1).

    Radioactive is the term used to describe isotopes which possess unstable nuclei that spontaneously emit energy (radiation) (section 17.1).

    Most stable isotopes have even numbers of neutrons or even numbers of protons, or both. There are very few stable isotopes which have odd numbers of neutrons and odd numbers of protons. It is the unstable nuclei which are said to be radioactive.

2. <u>Name and write the symbols that indicate the composition of the three types of radiation given off by naturally occurring radioactive materials</u> (section 17.3).

    An alpha particle is represented by the symbol $_2^4\alpha$. Such particles are composed of two protons and two neutrons, which give them a mass of 4 amu and a charge of +2.

    A beta particle is represented by the symbol $_{-1}^0\beta$. The charge on a beta particle is −1 and it has a mass of 0 amu.

    Gamma rays are not considered to be particles, but rather pure energy without mass or charge. They are very high energy radiation, somewhat like x-rays. The symbol for gamma rays is $_0^0\gamma$.

3. <u>Write balanced nuclear equations for various alpha and beta decay processes</u> (section 17.4).

    Alpha particle decay, the emission of an alpha particle from a nucleus, always results in the formation of an isotope of a different element. The product nucleus of such decay has an atomic number that is two less than that of the original nucleus and a mass number that is four less.

Example: Write a balanced nuclear equation for the alpha decay of $^{200}_{84}$Po.

Solution:

Step 1: First write the symbol of the radioactive isotope with an arrow to yield an alpha particle.

$$^{200}_{84}\text{Po} \longrightarrow {}^{4}_{2}\alpha$$

Step 2: Next subtract 4 from the mass number of the parent isotope (Po) and subtract 2 from the atomic number of the parent isotope to give the characteristics of the daughter nucleus.

$$^{200}_{84}\text{Po} \longrightarrow {}^{4}_{2}\alpha + {}^{196}_{82}$$

Step 3: Using the new atomic number (82), find the symbol of the new isotope (Pb) from the periodic table, and complete the equation.

$$^{200}_{84}\text{Po} \longrightarrow {}^{4}_{2}\alpha + {}^{196}_{82}\text{Pb}$$

As a final check, the sum of all the superscripts on the right in a nuclear equation must equal those on the left, and the sum of all the subscripts on the left must equal those on the right.

Beta decay also always results in the formation of an isotope of a different element. The mass number of the new isotope is the same as that of the original atom. The atomic number, however, has increased by one unit.

Example: Write a balanced nuclear equation for the beta decay of $^{25}_{11}$Na.

Solution:

Step 1: Write the symbol of the parent isotope and an arrow to yield a beta particle.

$$^{25}_{11}\text{Na} \longrightarrow {}^{0}_{-1}\beta$$

Step 2: To obtain the characteristics of the daughter nucleus, add one to the atomic number of the parent isotope and keep the mass number the same.

$$^{25}_{11}Na \longrightarrow \,^{\,\,\,0}_{-1}\beta + \,^{25}_{12}$$

Step 3: Determine the symbol of the new isotope from the periodic table and complete the equation.

$$^{25}_{11}Na \longrightarrow \,^{\,\,\,0}_{-1}\beta + \,^{25}_{12}Mg$$

To check the correctness of a beta decay, the algebraic sum of the subscripts on the right $12 + (-1) = 11$ must equal the subscript on the right.

4. <u>Indicate the changes that occur in the atomic number and mass number of a radioisotope as a result of alpha or beta decay</u> (section 17.4).

In alpha decay, the mass number of the daughter isotope is 4 less than that of the parent isotope, and the atomic number of the daughter isotope is 2 less than that of the parent isotope.

In beta decay, the mass number of the daughter isotope is the same as that of the parent isotope, and the atomic number of the daughter isotope is one more than that of the parent isotope.

5. <u>State the relative penetrating abilities of alpha, beta and gamma radiation</u> (section 17.5).

Alpha particles have low penetrating power and cannot penetrate the body's outer layers of skin.

Beta particles have greater penetrating power and can cause severe skin burns if their source remains in contact with the skin for an appreciable amount of time.

Gamma radiation has the greatest penetrating power and can penetrate deeply into organs, bone and tissue.

6. <u>Define the terms transmutation reaction and bombardment reaction</u> (section 17.6).

A transmutation reaction is a nuclear reaction in which one isotope is changed into an isotope of another element.

Radioactive decay is an example of a natural transmutation reaction.

A bombardment reaction is a nuclear reaction in which small particles traveling at very high speeds are collided with stable nuclei causing them to undergo nuclear change.

Example: $^{14}_{7}N + ^{4}_{2}\alpha \longrightarrow ^{17}_{8}O + ^{1}_{1}H$

In this bombardment reaction an alpha particle is collided with the nucleus of a stable nitrogen atom. The products are oxygen–17 and hydrogen–1. This reaction is also a transmutation reaction since isotopes of other elements are formed.

7. <u>Balance nuclear equations representing bombardment reactions</u> (section 17.6).

    Example: Write the equation for the bombardment reaction in which beryllium–9 captures an alpha particle and then emits a neutron.

    Solution:

        Step 1: The following information is given in the problem statement:

$$^{9}_{4}Be + ^{4}_{2}\alpha \longrightarrow ^{1}_{0}n +$$

        Step 2: Since the sum of the superscripts on each side of the equation must be equal, the superscript for the new isotope must be 12. In order for the sum of the subscripts on each side of the equation to be equal the subscript for the new isotope must be 6. Looking at the periodic table we determine the element with an atomic number of 6 is carbon (C). Therefore,

$$^{9}_{4}Be + ^{4}_{2}\alpha \longrightarrow ^{1}_{0}n + ^{12}_{6}C$$

A summary of notations used and the characteristics of the various types of radiation and small particles involved in transmutation reactions is given in Table 17–1 on page 459 of your textbook.

8. <u>Write balanced nuclear equations for various positron and electron capture decay processes</u> (section 17.6).

Positron decay involves the emission of a positron ($_{1}^{0}\beta$) from a synthetically produced radioisotope. The net effect of positron emission is to decrease the atomic number by one while the mass number remains constant. An example of a positron emission is:

$$_{15}^{30}P \longrightarrow \ _{1}^{0}\beta + _{14}^{30}S$$

Electron capture results from the pulling of an electron from a low energy orbital into the nucleus. An example of an electron capture is:

$$_{37}^{87}Rb + \ _{-1}^{0}e \longrightarrow \ _{36}^{87}Kr$$

The atomic number in an electron capture is decreased by one and the mass number remains the same.

9. <u>Indicate the changes that occur in the atomic number and mass number of a radioisotope as a result of positron emission or electron capture</u> (section 17.6).

In both positron emission and electron capture, the atomic number of the daughter isotope is one less than the parent isotope, and the mass number does not change.

10. <u>Define the term half-life and be able to calculate the fraction of a radioisotope left after a given whole number of half-lives have elapsed or vice versa</u> (section 17.7).

The half-life is the time required for one-half of any given quantity of a radioactive substance to undergo decay.

For example, if a radioisotope's half-life is 12 days and you have a 4.00 gram sample of it, then after 12 days (one half-life) only 2.00 grams of the sample (one-half the original amount) will remain undecayed.

Calculations involving amounts of radioactive material decayed, amounts remaining undecayed and time elapsed can be carried out using the following equation:

$$\begin{array}{ccc} \text{amount of} & & \\ \text{radioisotope} & & \text{original} \\ \text{undecayed} & = & \text{amount of} \quad \times \quad \dfrac{1}{2^{n}} \\ \text{after } n & & \text{radioisotope} \\ \text{half-lives} & & \end{array}$$

Examples 17-3 and 17-4 on page 463 of your textbook are

examples of the use of this equation.

11. <u>Relate nuclear stability to the total number of nucleons present and to the neutron/proton ratio</u> (section 17.8).

The relationship of the number of nucleons present to the nuclear stability is that every known isotope of every element with atomic number greater than 83 is unstable.

For elements of low atomic number, the neutron/proton ratios of stable isotopes are very close to one. For heavier elements, stable isotopes have higher neutron/proton ratios, with the ratio reaching approximately 1.5 for the heaviest stable elements.

Figure 17-5 on page 465 in your textbook shows graphically the relationship of the neutron/proton ratio to nuclear stability.

12. <u>Relate mode of decay for a radioisotope to neutron/proton ratio</u> (section 17.9).

Beta emission is the predominant decay mode for nuclides whose neutron/proton ratios are too high for stability.

Positron emission or electron capture are the modes of decay for nuclides whose neutron/proton ratios are too low for stability.

For radionuclei containing more than 209 nucleons usually more than one decay step is required to reach stability. Some decay steps involve alpha particle emission and others involve beta particle emission.

13. <u>State general methods of production for and stability characteristics of transuranium elements</u> (section 17.10).

Bombardment reactions are the source of the transuranium elements (elements 93 to 106).

All isotopes of all of the transuranium elements are radioactive. Information concerning the stability of the transuranium elements is given in Table 17-4 on page 468 of your textbook.

14. <u>Describe the general characteristics of a nuclear fission reaction and indicate how the fission process finds use in nuclear weapons and nuclear power plants</u> (section 17.11).

Nuclear fission is the process in which a heavy element

nucleus splits into two or more medium-sized nuclei as the result of bombardment. When fission occurs, very large amounts of energy are released, many times greater than from ordinary radioactive decay. This large amount of released energy, often called nuclear energy or atomic energy, is the most important aspect of the fission reaction.

In atomic bombs, uranium-235 undergoes fission when bombarded with slow-moving neutrons (thermal neutrons). There is no unique way in which the uranium-235 splits, and more than 200 nuclides of 35 different elements have been identified as products of the reaction. In nuclear weapons, the fission reaction is uncontrolled.

In nuclear power plants, the uranium is present in the reactor core in the form of pellets of the oxide U_3O_8 enclosed in long steel tubes. The uranium is a mixture of all isotopes of uranium, not just uranium-235, due to the cost of isotope separation. The energy from fission appears as heat, which is drawn out of the reactor core to turn water into steam which is used to generate electricity. In nuclear power plants, the fission reaction is carried out under controlled conditions.

15. <u>Describe the general characteristics of a nuclear fusion reaction and indicate the types of problems which must be overcome if this process is to be used in a practical manner</u> (section 17.12).

Nuclear fusion is the putting together of small atoms to make larger ones. It is, thus, essentially the opposite of nuclear fission.

The problems which must be overcome if nuclear fusion is to be used in a practical manner are containment of the reaction mixture and development of a method of heating the reaction mixture to about 100 million degrees in order to start the fusion reaction.

16. <u>Contrast the major differences between nuclear reactions and "ordinary" chemical reactions</u> (section 17.13).

Table 17-5 on page 474 of your textbook compares the differences between nuclear reactions and "ordinary" chemical reactions.

1. Complete the following table:

Name	Symbol	Unit Charge	Mass amu	Relative Penetrating Power
Alpha Particle				
Beta Particle				
Gamma Ray				

2. Complete and balance the following nuclear equations for alpha or beta decay:

a) $^{32}_{14}Si \longrightarrow \ ^{0}_{-1}\beta \ + \ \underline{\hspace{2cm}}$

b) $^{240}_{96}Cm \longrightarrow \ ^{4}_{2}\alpha \ + \ \underline{\hspace{2cm}}$

c) $^{87}_{36}Kr \longrightarrow \ ^{0}_{-1}\beta \ + \ \underline{\hspace{2cm}}$

d) $^{232}_{92}U \longrightarrow \ ^{4}_{2}\alpha \ + \ \underline{\hspace{2cm}}$

e) $^{208}_{84}Po \longrightarrow \ ^{4}_{2}\alpha \ + \ \underline{\hspace{2cm}}$

f) $^{66}_{29}Cu \longrightarrow \ ^{0}_{-1}\beta \ + \ \underline{\hspace{2cm}}$

3. Complete and balance the following nuclear equations for bombardment reactions:

a) $^{14}_{7}N + ^{1}_{1}P \longrightarrow \underline{\hspace{2cm}} + ^{4}_{2}\alpha$

b) $^{27}_{13}Al + ^{1}_{0}n \longrightarrow \underline{\hspace{2cm}} + ^{1}_{1}P$

c) $^{63}_{29}Cu + ^{2}_{1}H \longrightarrow \underline{\hspace{2cm}} + ^{3}_{1}H$

d) $^{63}_{29}Cu + \underline{\hspace{2cm}} \longrightarrow ^{61}_{28}Ni + ^{4}_{2}\alpha$

e) $^{44}_{20}Ca + ^{1}_{1}P \longrightarrow \underline{\hspace{2cm}} + ^{1}_{0}n$

f) $^{4}_{2}He + \underline{\hspace{2cm}} \longrightarrow ^{12}_{6}C + ^{1}_{0}n$

4. Complete and balance the following nuclear equations for positron emission and electron capture decay processes:

a) $^{140}_{59}Pr \longrightarrow \ ^{0}_{1}\beta \ + \ \underline{\hspace{2cm}}$

b) $^{18}_{10}Ne \longrightarrow \ ^{0}_{1}\beta \ + \ \underline{\hspace{2cm}}$

c) $^{7}_{4}Be + ^{0}_{-1}e \longrightarrow$ _____

d) _____ $+ ^{0}_{-1}e \longrightarrow ^{37}_{17}Cl$

e) $^{109}_{49}In \longrightarrow ^{0}_{1}\beta +$ _____

f) $^{161}_{69}Tm + ^{0}_{-1}e \longrightarrow$ _____

5. If 4.00 g of a radioactive sample has a half-life of 73 hours, how many grams will remain undecayed after 6 days and 2 hours (146 hours)?

6. A radioactive isotope has a half-life of 1 minute. If 5.00 grams decays over a period of 4 minutes, how many grams will remain undecayed?

7. If 1.20 g of a radioactive isotope decays to 0.30 g in 40 minutes, what is its half-life?

8. What is the half-life of 0.80 g of a radioactive isotope, which decays to 0.10 g in 90 seconds?

9. Nuclear fusion involves the putting together of _____ nuclei to produce _____ nuclei.

10. Nuclear fission involves the bombardment of _____ nuclei to produce _____ nuclei.

MULTIPLE CHOICE EXERCISES

1. The three types of radiation emitted by naturally occurring radioactive elements are:

    a) electrons, protons and neutrons
    b) x-rays, gamma rays and delta rays
    c) alpha particles, beta particles and gamma rays
    d) alpha particles, gamma rays and neutrons

2. Which of the following types of radiation is composed of particles which carry a -1 charge?

    a) alpha
    b) beta
    c) gamma
    d) delta

3. Alpha particles are composed of:

    a) protons and electrons
    b) protons and neutrons
    c) neutrons and electrons
    d) protons, neutrons and electrons

4. Which of the following is not a balanced nuclear equation?

    a) $^{121}_{50}Sn \longrightarrow ^{\ 0}_{-1}\beta + ^{121}_{51}Sb$

    b) $^{238}_{92}U \longrightarrow ^{4}_{2}\alpha + ^{234}_{90}Th$

    c) $^{10}_{4}Be \longrightarrow ^{10}_{3}Li + ^{\ 0}_{-1}\beta$

    d) $^{190}_{78}Pt \longrightarrow ^{186}_{76}Os + ^{4}_{2}\alpha$

5. When the equation is balanced, the missing product is:

    $$^{211}_{83}B \longrightarrow ^{207}_{81}Tl + \underline{\hspace{2cm}}$$

    a) $^{4}_{2}\alpha$

    b) $^{2}_{1}H$

    c) $^{4}_{3}Li$

    d) $^{4}_{2}Li$

6. The missing product in the reaction is:

    $$^{67}_{30}Zn + ^{1}_{0}n \longrightarrow ^{4}_{2}\alpha + \underline{\hspace{2cm}}$$

    a) $^{68}_{28}Ni$

    b) $^{70}_{32}Ge$

    c) $^{64}_{30}Zn$

    d) $^{64}_{28}Zn$

7. The mass number remains constant during which of the following modes of decay?

    a) alpha emission
    b) both alpha and beta emission
    c) both alpha and gamma emission
    d) both beta and gamma emission

8. The explanation for how a beta particle is produced in the nucleus and then ejected involves the conversion (in a complex series of steps) of a:

    a) proton to a neutron
    b) neutron to a proton
    c) proton to an electron
    d) neutron to an electron

9. Which of the following is an incorrect statement concerning bombardment reactions?

    a) new elements not found in nature have been produced
    b) at least one radioisotope of every known element has been produced
    c) unstable nuclides have been changed into stable nuclides
    d) elements are transmuted into other elements

10. Alpha and beta particles and gamma rays are called ionizing radiation because as they travel through matter they:

    a) attract ions
    b) repel ions
    c) knock electrons off atoms or molecules in their path
    d) decay into ions

11. Which of the following would have the greatest penetrating power into matter?

    a) alpha particles
    b) gamma rays
    c) beta particles
    d) protons

12. Positron emission is a mode of decay found:

    a) only among naturally occurring radioisotopes
    b) only among synthetic radioisotopes
    c) among both naturally occurring and synthetic isotopes
    d) only among transuranium elements

13. The isotope $^{56}_{25}$Mn spontaneously decays to give $^{56}_{26}$Fe. The process involves the:

    a) capture of an electron by the Mn-56 nucleus
    b) emission of a positron
    c) emission of an alpha particle
    d) emission of a beta particle

14. A radioisotope has a half-life of 10 years. If you have a 4 gram sample of this radioisotope today, how many grams would you expect to have in 30 years?

    a) 1/2 gram
    b) 1 gram
    c) 2 grams
    d) 8 grams

15. If a 2 gram sample of a radioactive substance decays to 1/4 gram of the substance in 60 minutes, the half-life of the radioactive substance is:

    a) 1/4 minute
    b) 20 minutes
    c) 30 minutes
    d) 180 minutes

16. A radioisotope which has too high of a neutron/proton ratio gains stability by:

    a) beta emission
    b) alpha emission
    c) gamma emission
    d) electron capture

17. For stable nuclei, as the size of the nucleus increases the neutron to proton ratio:

    a) increases
    b) decreases
    c) remains constant
    d) fluctuates wildly

18. Which of the following statements concerning synthetic elements is incorrect?

    a) they are produced via bombardment reactions
    b) significant quantities of only a few elements have been produced
    c) stable isotopes of some of them have been produced
    d) a total of 18 synthetic elements have been produced

19. Which of the following processes is directly related to the concept "splitting the atom?"

    a) electron capture
    b) nuclear fission
    c) nuclear fusion
    d) positron decay

20. Which of the following statements concerning ^{235}U is correct?

   a) ^{235}U nuclei always fragment to give nuclei of equal mass

   b) ^{235}U fission always produces isotopes of the same two elements

   c) there is no unique way in which the ^{235}U nucleus fragments

   d) two neutrons are always produced as the result of ^{235}U fission

21. A fourth state of matter, called a "plasma" is very important in:

   a) nuclear fusion
   b) nuclear fission
   c) breeder reactors
   d) bombardment reactions

22. The source of $^{2}_{1}H$ (deuterium) for use in fusion processes is:

   a) deuterium ore
   b) sea water
   c) the atmosphere
   d) burnt pizza crust

ANSWERS TO PROBLEMS SET

1.

Name	Symbol	Unit Charge	Mass amu	Relative Penetration Power
Alpha Particle	α	+2	4	low
Beta Particle	β	-1	0	moderate
Gamma Ray	γ	0	0	high

2. a) $^{32}_{15}P$ d) $^{228}_{90}Th$

   b) $^{236}_{94}Pu$ e) $^{204}_{82}Pb$

   c) $^{87}_{37}Rb$ f) $^{66}_{30}Zn$

3. a) $^{11}_{6}B$

   b) $^{27}_{12}Mg$

   c) $^{62}_{29}Cu$

   d) $^{2}_{1}H$

   e) $^{44}_{21}Sc$

   f) $^{9}_{4}Be$

4. a) $^{140}_{58}Ce$

   b) $^{18}_{9}F$

   c) $^{7}_{3}Li$

   d) $^{37}_{18}Ar$

   e) $^{109}_{48}Cd$

   f) $^{161}_{68}Tm$

5. 1.00 g

6. 0.675 g

7. 20 minutes

8. 30 seconds

9. Smaller, larger

10. Larger, medium-sized

ANSWER TO MULTIPLE CHOICE EXERCISES

1. c

2. b

3. b

4. c

5. a

6. d

7. d

8. b

9. c

10. c

11. b

12. b

13. d

14. a

15. b

16. a

17. a

18. c

19. b

20. c

21. a

22. b

CHAPTER 18
Introduction to Organic Chemistry

REVIEW OF CHAPTER OBJECTIVES

1. Distinguish between organic and inorganic chemistry in terms of modern day definitions, elements studied and number of compounds (section 18.1).

   Organic chemistry is defined today as the study of hydrocarbons (binary compounds of hydrogen and carbon) and their derivatives.

   In a less rigorous manner, organic chemistry is often defined as the study of carbon-containing compounds. Using this classification, all carbon-containing compounds are considered organic except the oxides of carbon, carbonates, cyanides and metallic carbides. These compounds are all considered to be inorganic rather than organic.

   Inorganic chemistry is the study of the compounds containing the other 105 elements, plus the exceptions just mentioned.

   Since carbon possesses the unique ability to bond to itself in long chains, rings and complex combinations of both, approximately 4 million organic compounds are known. The other 105 elements together form only one-quarter million compounds. The existence of 15 organic compounds for every one inorganic compound is the reason the study of carbon-containing compounds is a field by itself.

2. Define the terms alkane, unsaturated hydrocarbon and homologous series (section 18.3).

   An alkane is a compound containing only carbon and hydrogen atoms held together by single bonds. Since each carbon atom in an alkane molecule is bonded to four other atoms, the maximum number possible, alkanes are often referred to as saturated hydrocarbons.

A homologous series is a group of compounds, each of which differs from the previous one in the series by a constant amount.

3. <u>Define the term structural isomerism and give examples of such isomerism in the alkane hydrocarbon series</u> (section 18.4).

Structural isomerism is the existance of compounds which have the same molecular formula, but different structural formulas.

Example: The structural formulas for the two butane isomers which exist are:

$$CH_3-CH_2-CH_2-CH_3 \qquad \text{butane}$$

$$CH_3-CH-CH_3$$
$$\qquad | $$
$$\qquad CH_3 \qquad \text{isobutane}$$

Both of these molecules have the molecular formula C_4H_{10}, but you can see that in butane the carbon atoms are all arranged in a straight chain, while in the isobutane there is a $-CH_3$ group attached to the center atom of a three-carbon chain.

4. <u>Given the structural formula for an alkane or cycloalkane, name it using IUPAC rules and vice versa</u> (sections 18.5 and 18.6).

To be able to successfully perform this objective, you must have a complete understanding of, and be able to apply, the IUPAC rules found on pages 485 through 488 in your textbook. You should memorize the following list of names for the first ten straight chain alkanes since they are the basis for the entire IUPAC system of nomenclature.

CH_4	methane
C_2H_6	ethane
C_3H_8	propane
C_4H_{10}	butane
C_5H_{12}	pentane
C_6H_{14}	hexane

C_7H_{16} heptane

C_8H_{18} octane

C_9H_{20} nonane

$C_{10}H_{22}$ decane

You should memorize the names of and be able to recognize the following alkyl groups when attached to the longest continuous chain:

CH_3——   methyl

CH_3-CH_2——   ethyl

$CH_3-CH_2-CH_2$——   propyl

$$\begin{array}{c} CH_3 \\ | \\ HC—— \\ | \\ CH_3 \end{array}$$ isopropyl

In the above alkyl groups the dash (——) which is not attached to anything denotes the point at which they would be attached to the longest continuous chain.

Example: Give the correct IUPAC name for the following compound:

$$\begin{array}{c} CH_3 \\ | \\ CH_3 \quad\quad CH_3-CH \\ | \quad\quad\quad\quad\quad | \\ CH_2-CH_2-CH-CH-CH-CH_3 \\ | \quad\quad\quad | \\ CH_2 \quad\quad CH_2-CH-CH_3 \\ | \quad\quad\quad\quad | \\ CH_3 \quad\quad\quad CH_3 \end{array}$$

Solution:

Step 1:   The longest continuous chain is:

$$\begin{array}{c} CH_3 \\ | \\ CH_2-CH_2-CH-CH-CH \\ | \\ CH_2-CH-CH_3 \end{array}$$

which contains 9 carbons. Thus, the base name will be nonane. Notice that the longest continuous chain does not have to be in a straight line.

Step 2: When the longest continuous chain is numbered from the left, the first alkyl group we come to is on carbon 4, while numbered from the right to left, the first alkyl group is on carbon 2. Therefore, the chain is correctly numbered starting from the right.

$$9 \; \overset{|}{CH_3} \qquad\qquad 4$$

$$\underset{8}{CH_2}-\underset{7}{CH_2}-\underset{6}{CH}-\underset{5}{CH}-\overset{}{CH}$$

$$\underset{3}{CH_2}-\underset{2}{CH}-\underset{1}{CH_3}$$

Step 3: The names of the alkyl groups and the number of the carbon to which they are attached are:

methyl on carbon number 2
methyl on carbon number 4
isopropyl on carbon number 5
ethyl on carbon number 6

The name of the compound is:

6-ethyl-2,4-dimethyl-5-isopropylnonane.

The name is written as one word with a hyphen used between the number (location) and name of the alkyl group. There is no space between the name of the last alkyl group (isopropyl) and the base name (nonane).

Step 4: The 2 methyl groups present are indicated by the prefix di- and their location on the longest continuous chain indicated by a number. These position numbers are separated by commas and put just before the numerical prefix, with hyphens before and after the numbers (-2,4-dimethyl-).

Note that the prefix must be accompanied by the quantity of numbers indicated by that prefix, even if the same number must be written twice, as in 3,3-dimethylpentane. Both threes must be written.

Step 5: Note that ethyl is named first due to the alphabetical rule determining the order in which alkyl groups are listed even though it has a larger number than any other group. Also note that prefixes do not affect the alphabetical order for alkyl groups; "e" from ethyl comes before "m" from methyl.

Structural formulas can easily be obtained from correct IUPAC names if you understand the IUPAC system for naming compounds.

Example: Draw the structural formula for

3,3-diethyl-2,4,6-triethyl-5-isopropyl-4-propylheptane

Solution:

Step 1: The IUPAC name indicates that the base chain contains 7 carbon atoms (heptane). Draw a heptane skeleton and number it. When drawing structures, it is easiest to always number it from left to right.

$$C-C-C-C-C-C-C$$
$$1 \ 2 \ 3 \ 4 \ 5 \ 6 \ 7$$

Step 2: Place two ethyl groups on carbon number 3.

$$
\begin{array}{c}
CH_3 \\
| \\
CH_2 \\
| \\
C-C-C-C-C-C-C \\
| \\
CH_2 \\
| \\
CH_3
\end{array}
$$

Step 3: Place methyl groups on carbons 2, 4 and 6.

$$
\begin{array}{c}
\qquad \quad CH_3 \\
\qquad \quad | \\
CH_3 \ CH_2 \ CH_3 \qquad CH_3 \\
| \quad | \quad | \qquad \quad | \\
C - C - C - C - C - C - C \\
\qquad \quad | \\
\qquad \quad CH_2 \\
\qquad \quad | \\
\qquad \quad CH_3
\end{array}
$$

Step 4: Place an isopropyl group on carbon number 5.

CH$_3$
|
CH$_3$ CH$_2$ CH$_3$ CH$_3$
| | | |
C – C – C – C – C – C – C
      | |
     CH$_2$ CH–CH$_3$
      | |
     CH$_3$ CH$_3$

Step 5:   Place a propyl group on carbon number 4.

CH$_3$
|
CH$_3$ CH$_2$ CH$_3$ CH$_3$
| | | |
C – C – C – C – C – C – C
      | | |
     CH$_2$ CH$_2$ CH–CH$_3$
      | | |
     CH$_3$ CH$_2$ CH$_3$
          |
        CH$_2$

Step 6:   Add the necessary hydrogens to the carbon base chain so that each carbon has four bonds.

CH$_3$
|
CH$_3$ CH$_2$ CH$_3$ CH$_3$
| | | |
CH$_3$ – CH – C – C – CH – CH – CH$_3$
             | | |
          CH$_2$ CH$_2$ CH – CH$_3$
           | | |
         CH$_3$ CH$_2$ CH$_3$
              |
           CH$_3$

Naming cycloalkanes is very similar to that for noncyclic alkanes with the addition of the following two changes to the IUPAC rules.

1. The prefix cyclo- is placed before the name that corresponds to the noncyclic chain that has the same number of carbon atoms as the ring.

2. Alkyl groups, when more than one is present, are located by numbering the carbons in the ring using a numbering system that gives the lowest numbers for the carbons at which the alkyl groups are attached.

Example: Give the IUPAC name for each of the following cycloalkanes:

a) <!-- cyclohexane with CH3 CH3 --> —CH$_3$ —CH$_3$

c) 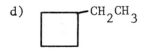 CH$_3$ CH$_3$

b) CH$_3$ —CH$_3$

d) —CH$_2$CH$_3$

Solution:

a) **Step 1:** The ring contains six carbons so it is named cyclohexane.

CH$_3$
CH$_3$

Step 2: The ring is numbered to give the lowest numbers for the carbons to which the alkyl groups are attached. The correct name is, thus,

1,2-dimethylcyclohexane

When numbering the ring always start at one of the carbons to which an alkyl group is attached. It is not critical which one you start on as long as you obtain the lowest possible numbers.

b) **Step 1:** The ring contains six carbons so it is a cyclohexane.

Step 2: CH$_3$ CH$_3$

The lowest set of numbers for the two methyl groups is 1 and 3. Thus, we have

1,3-dimethylcyclohexane

c) **Step 1 :** The ring contains six carbons so it is a cyclohexane.

Step 2: There are two methyl groups on the same

carbon so it is named

1,1-dimethylcyclohexane

d) **Step 1:**   The ring contains four carbons so it is a cyclobutane.

   Step 2:   There is an ethyl group attached to the ring so it is named

   ethylcyclobutane.

   Notice when only one alkyl group is attached it is not necessary to number the ring since whichever carbon it is on will be carbon number 1.

To draw the structure of a cycloalkane, the process is the same as for noncyclic alkanes.

5. <u>Define the terms unsaturated hydrocarbon, alkene and alkyne</u> (section 18.7).

An unsaturated hydrocarbon is a hydrocarbon that contains fewer hydrogen atoms than the maximum possible for the number of carbon atoms which a molecule contains. Unsaturation results from the presence of one or more multiple bonds.

Ethane (C_2H_6) is called a saturated hydrocarbon since it contains the maximum number of hydrogen atoms (6) possible for a molecule possessing 2 carbon atoms.

Ethene (C_2H_4) and ethyne (C_2H_2) are called unsaturated hydrocarbons since they contain less than the maximum number of hydrogen atoms (6) as found in ethane.

Alkenes are hydrocarbons in which there is one carbon-carbon double bond per molecule. Ethene (C_2H_4) is the first member of the alkene series.

Alkynes are hydrocarbons in which there is one carbon-carbon triple bond per molecule. Ethyne (C_2H_2) is the first member of the alkyne series.

6. <u>Given the structural formula for an unsaturated hydrocarbon (an alkene or alkyne), name it using IUPAC rules, and vice versa</u> (section 18.7).

The rules for naming alkenes and alkynes are the same as those for naming alkanes with the following changes.

1. The -ane ending characteristic of alkanes is changed to -ene for alkenes and to -yne for alkynes.

2. For noncyclic molecules containing more than 3 carbon atoms, the position of the multiple bond is indicated by a single number placed in front of the base chain name corresponding to the lower numbered carbon involved in the multiple bond. The chain is always numbered in such a way as to give the number assigning the position of the multiple bond the lowest number possible.

Example: Name the following unsaturated hydrocarbons:

a) $CH_3-CH=CH_2$

b) $CH_3-\underset{\underset{\displaystyle CH_3}{|}}{C}=CH-CH_3$

c) $CH_3-C\equiv C-\underset{\underset{\displaystyle CH_3}{|}}{CH}-CH_3$

d) $CH_3-CH=C=CH_2$

Solution:

a) The chain contains 3 carbons; therefore, the name is propene. Notice that the -ane ending used for an unsaturated hydrocarbon has been changed to -ene to indicate this molecule contains a double bond. No number is necessary for this compound since both positions where the double bond could be placed are equivalent.

b) The longest continuous chain contains 4 carbon atoms; therefore, the base name is butene. Numbering from the left gives the IUPAC name 2-methyl-2-butene. The 2 in front of the methyl group tells us it is located on carbon 2 and the 2 in front of the butene indicates that the double bond is located between carbon atoms 2 and 3. If this compound were incorrectly named 3-methyl-2-butene, the double bond would still have the lowest number, but the methyl group would not have the lowest possible number.

c) The longest continuous chain containing the triple bond contains 5 carbon atoms; therefore, it is a pentyne. The triple bond is between carbons 2 and 3 so it is a 2-pentyne. Since the

triple bond is between carbons 2 and 3, the methyl
group is on carbon number 4 and the correct IUPAC
name is 4-methyl-2-pentyne.

d) The chain containing the two double bonds
contains 4 carbons; therefore, it is a butadiene.
When a chain contains more than one multiple bond
the number of multiple bonds is indicated by a
Greek prefix just as more than one of the same
alkyl groups on a chain are indicated using Greek
prefixes. The positions of the two double bonds
are indicated by the lowest possible combination
of numbers. The correct IUPAC name is 1,2-
butadiene.

Structural formulas for unsaturated hydrocarbons are
obtained from the correct IUPAC names using the same
procedures as are used for saturated alkanes.

Example: Draw the structural formulas for the following:

  a) 3,4-dimethyl-1-pentene
  b) 5-methyl-2-hexyne
  c) 1,3-cyclohexadiene

Solution:

a) **Step 1:** The IUPAC name indicates that the base
chain contains 5 carbons and a double
bond (indicated by the -ene ending).
The double bond is located between
carbons 1 and 2.

C=C-C-C-C

Step 2: Place a methyl group on carbons 3 and 4.

$$CH_3 \quad CH_3$$
$$| \qquad |$$
$$C = C - C - C - C$$

Step 3: Add the necessary hydrogens to the
carbon base chain so that each carbon
has 4 bonds.

$$CH_3 \quad CH_3$$
$$| \qquad |$$
$$CH_2 = CH - CH - CH - CH_3$$

Notice that the first carbon on the
right only has two hydrogen atoms

attached even though it is at the end of the chain. This is due to the fact that 2 of that carbon's 4 possible bonds are used in the double bond formation. The second carbon in the chain has only one hydrogen instead of the customary 2 for the same reason.

b) **Step 1:** The IUPAC name indicates that the base chain contains 6 carbons and a triple bond (indicated by the -yne ending). The triple bond is located between carbons 2 and 3.

$$C - C \equiv C - C - C - C$$

Step 2: Place a methyl group on carbon number 5.

$$\overset{\overset{\displaystyle CH_3}{\displaystyle |}}{C - C = C - C - C - C}$$

Step 3: Add the necessary hydrogens to the carbon base chain so that each carbon has 4 bonds.

$$CH_3 - C \equiv C - CH_2 - \overset{\overset{\displaystyle CH_3}{\displaystyle |}}{CH} - CH_3$$

Notice that carbons 3 and 4 contain no hydrogens since all 4 of their possible bonds are used in forming the triple bond and bonding to another carbon in the chain.

c) **Step 1:** The IUPAC name indicates 6 carbons in a ring containing two double bonds. One double bond is between carbons 1 and 2 and the other between carbons 3 and 4.

This is the complete structure since the symbols for all the carbons and hydrogens are understood in a cycloalkene, just as for cycloalkanes.

7. <u>Describe how the bonding in aromatic hydrocarbons differs from that in nonaromatic hydrocarbons</u> (section 18.8).

The bonding in aromatic hydrocarbons involves delocalized bonding, while nonaromatic hydrocarbon bonding involves only localized bonds. Delocalized bonding involves the sharing of the same valence electrons by three or more atoms, while in localized bonding the valence electrons are shared by only two atoms. For a complete discussion of the bonding in the benzene molecule, see pages 495 to 497 in your textbook.

8. <u>Name, using IUPAC or common names, the simpler aromatic hydrocarbons</u> (section 18.8).

The first member of the aromatic hydrocarbon homologous series is benzene. Following are the IUPAC nomenclature rules for naming substituted aromatic hydrocarbons.

1. If there is only one alkyl group on the ring, the name of the alkyl group is attached to the name benzene and it is written as one word.

Examples:

    methylbenzene   ethylbenzene

2. If two or more alkyl groups are attached to the ring, the carbons in the ring are numbered in such a manner that the lowest combination of numbers is obtained. Greek prefixes are still used if the 2 alkyl groups are the same.

a)

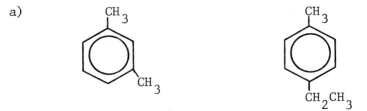

  1,3-dimethylbenzene   1-ethyl-4-methylbenzene

The names of the alkyl groups are placed in alphabetical order when naming substituted aromatic hydrocarbons. Since the ethyl group is named first, the carbon to which it is attached is numbered 1 for simplicity.

A common system of nomenclature for disubstituted

aromatic rings involves the use of the prefixes ortho (o-) for 1,2 substitution, meta (m-) for 1,3 substitution and para (p-) for 1,4-substitution.

Examples:

m-ethylmethylbenzene   o-dimethylbenzene

9. List the major sources and uses of hydrocarbons (section 18.9).

The major sources and uses are given in Table 18-3 on page 499 of your textbook.

10. Define the terms hydrocarbon derivative and functional group and know the meaning associated with the symbol R (section 18.10).

A hydrocarbon derivative is a hydrocarbon molecule in which one or more of the hydrogen atoms has been replaced with a new atom or group of atoms.

Example: When one of the hydrogen atoms of methane (CH_4) is replaced with a hydroxyl group, we have methyl alcohol (CH_3OH). Methyl alcohol is considered a derivative of the hydrocarbon methane.

A functional group is that part of a hydrocarbon derivative that contains the elements other than carbon or hydrogen.

In the preceding example, the hydroxyl group (OH) is considered a functional group.

The symbol R is used in general formulas to designate the hydrocarbon part for a class of hydrocarbon derivatives all of which possess the same functional group (see page 500 in your textbook).

11. Know the general structural features, including functional group, associated with the following classes of hydrocarbon derivatives: halogenated hydrocarbons, alcohols, ethers, carboxylic acids and esters (sections 18.11 through 18.15).

Halogenated hydrocarbons (or organic halides) are
hydrocarbon derivatives in which one or more hydrogen
atoms in the parent hydrocarbon have been replaced by a
halogen atom. Halogenated hydrocarbons have the general
formula R–X, where R is the hydrocarbon part of the
molecule and X represents any of the halogens fluorine,
chlorine, bromine or iodine.

Alcohols are hydrocarbon derivatives in which one or more
of the hydrogen atoms in the parent hydrocarbon have been
replaced by hydroxyl groups (–OH groups). Alcohols have
the general formula R–OH, where R represents the
hydrocarbon part of the molecule.

Ethers are compounds with the general formula R–O–R.
The R groups may or may not be the same.

Carboxylic acids are compounds which contain the carboxyl
functional group ($-\overset{\overset{\text{O}}{\|}}{\text{C}}$–OH). Carboxylic acids have the
general formula R–$\overset{\overset{\text{O}}{\|}}{\text{C}}$–OH, where R represents the
hydrocarbon part of the molecule.

Esters are compounds which contain the $-\overset{\overset{\text{O}}{\|}}{\text{C}}$–O–R functional
group. Esters have the general formula R–$\overset{\overset{\text{O}}{\|}}{\text{C}}$–O–R where the
R's represent the hydrocarbon parts of the molecule. The
R's may be the same or different. Esters are very
closely related to carboxylic acids and may be thought of
as a carboxylic acid in which an R group has replaced the
hydrogen atom attached to the oxygen in the acid.

12. Be able to give both IUPAC and common names for the
simpler halogenated hydrocarbons, alcohols, ethers,
carboxylic acids and esters (sections 18.11 through
18.15).

In the IUPAC system for naming halogenated hydrocarbons
the prefixes fluoro- for F, chloro- for Cl, bromo- for
Br and iodo- for I are used to designate the substituted
halogen atom. These prefixes with location numbers, if
isomers are possible, are attached to the name of the
longest continuous chain. Thus,

$$\underset{\overset{\displaystyle|}{\text{Cl}}}{\text{CH}_3\text{–CH}}\text{–CH}_2\text{–CH}_2\text{–CH}_3$$

would be named 2-chloropentane.

The IUPAC name of an alcohol containing only one hydroxyl group is obtained from the name of the parent alkane by replacing the -e ending with -ol. When necessary, the position of the -OH functional group is indicated by a number. Thus,

$$CH_3-\underset{\underset{\displaystyle OH}{|}}{CH}-CH_3$$

would be named 2-propanol. The simpler alcohols are often known by common names, where the alkyl group name is followed by the word alcohol. Thus, 2-propanol is also known by the common name isopropyl alcohol.

IUPAC names for ethers are obtained by calling the -O-R functional group an alkoxy group. The alkoxy group is then named as a substituent on the parent hydrocarbon molecule. In mixed ethers, the smallest alkyl group becomes the alkoxy group. Thus,

$$CH_3\underset{\underset{\displaystyle O-CH_3}{|}}{CH}-CH_3$$

would be named 2-methoxypropane. Common names for ethers are obtained by first naming the two alkyl groups and adding the word ether. Thus, 2-methoxypropane would have the common name methyl propyl ether.

IUPAC names for carboxylic acids are derived by replacing the -e of the name of the longest carbon chain containing the carboxyl group ($-C\overset{\displaystyle O}{\underset{\displaystyle OH}{\diagdown}}$) with -oic and then adding the word acid. The functional group carbon atom is counted as part of the longest chain and is assigned the number 1 when a numbering system is needed. Thus,

$$CH_3\underset{\underset{\displaystyle CH_3}{|}}{CH}-C\overset{\displaystyle O}{\underset{\displaystyle OH}{\diagdown}}$$

would be called 2-methylpropanoic acid.

Names of esters bear a direct relation to the names of the acids from which they may be considered to be derived. The IUPAC name of the ester is formed by changing the -ic ending of the acid name (either common or IUPAC name) to -ate and preceding this name with the name of the R group attached to the oxygen atom. Thus,

$$\overset{\overset{\displaystyle O}{\displaystyle \|}}{CH_3-C}=O-CH_2CH_3$$

is named ethylethanoate using the IUPAC system and ethyl acetate using the common system.

13. <u>List uses for selected halogenated hydrocarbons, alcohols, ethers, carboxylic acids and esters</u> (sections 18.11 through 18.15).

The uses of many commercially important halogenated hydrocarbons are listed in Table 18-5 on page 502 of your textbook.

The uses of common alcohols are given on pages 504 and 505 of your textbook.

Ethers are primarily used as anesthetics.

Commonly encountered carboxylic acids and their uses are given in Table 18-8 on page 508 of your textbook.

Esters are primarily used as artificial flavoring agents. Table 18-9 on page 509 of your textbook lists some of the common ester flavoring agents and the tastes associated with them.

PROBLEM SET

1. Draw and name all 5 carbon alkane isomers.

2. Using the IUPAC system, name the following alkanes:

a) $CH_3-CH_2-\underset{\underset{\displaystyle CH_3}{\displaystyle |}}{\underset{\underset{\displaystyle CH_2}{\displaystyle |}}{CH}}-CH_2-CH_3$

b) $CH_3-\underset{\underset{\displaystyle CH_3}{\displaystyle |}}{\overset{\overset{\displaystyle CH_3}{\displaystyle |}}{C}}-CH_3$

c) $CH_3-\underset{\underset{\displaystyle CH_3}{\displaystyle |}}{\overset{\overset{\displaystyle CH_3}{\displaystyle |}}{CH}}-CH_2-CH_2-CH_2-\underset{\underset{\displaystyle CH_2}{\displaystyle |}}{\overset{\overset{\displaystyle CH_3}{\displaystyle |}}{CH}}-CH_3$ $\underset{CH_3}{|}$

d) $CH_3-\underset{\underset{\displaystyle CH_3}{\displaystyle |}}{\underset{\underset{\displaystyle CH_2}{\displaystyle |}}{CH}}-CH_2-\overset{\overset{\displaystyle CH_3}{\displaystyle |}}{CH}-CH_2-\underset{\underset{\displaystyle CH_3}{\displaystyle |}}{CH_2}$

3. Draw the structural formula for each of the following compounds:

   a) 2-methylhexane
   b) 2,3-dimethylbutane

   c) 3,3-dimethyl-4-ethylhexane
   d) 2,2,4-trimethylpentane

4. Using the IUPAC system, name the following cycloalkanes:

a)

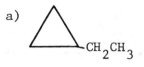

b)

c)

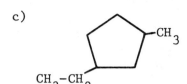

d)

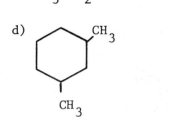

5. Write structural formulas for each of the following cycloalkanes:

   a) propylcyclopentane
   b) isopropylcyclohexane
   c) 1,3-dimethyl-2-ethylcyclohexane
   d) 1,1-dimethylcyclopropane

6. Using the IUPAC system, name each of the following unsaturated hydrocarbons:

a)

$$CH_3-\overset{\overset{\displaystyle CH_3}{|}}{C}=CH-CH_3$$

b)

$$CH_2=CH-\overset{\overset{\displaystyle CH_3}{|}}{CH}-CH_3$$

c) $CH_2=CH-CH=CH_2$

d) $CH_3-C\equiv C-CH_2-CH_3$

e)

$$CH_3-\overset{\overset{\displaystyle CH_3}{|}}{\underset{\underset{\displaystyle CH_3}{|}}{C}}=C-CH_3$$

f)

7. Write structural formulas for the following compounds:

a) 1-butene
b) 4-ethyl-2-heptene
c) 3-methyl-1-pentyne

d) 4,4,5-trimethyl-2-hexyne
e) 1,3-pentadiene
f) 2-methyl-1,3-cyclohexadiene

8. Give the IUPAC name for each of the following aromatic hydrocarbons:

a)

d)

b)

e)

c)

f)

9. Write structural formulas for the following aromatic hydrocarbons:

a) 1,4-dimethylbenzene
b) p-xylene
c) m-dimethylbenzene

d) 1-ethyl-2-propylbenzene
e) 1,3-dimethylbenzene
f) 1-methyl-2-ethylbenzene

10. What kind of hydrocarbon derivative is each of the following:

a) CH_3-CH_2Br

b) CH_3-OH

c) CH_3-O-CH_3

d) $H-\overset{\overset{\displaystyle O}{\|}}{C}-OH$

e) $CH_3-O-\overset{\overset{\displaystyle O}{\|}}{C}-H$

11. Name the following halogenated hydrocarbons using IUPAC rules:

a) $CH_3-\underset{\underset{Br}{|}}{CH}-CH_2-CH_3$

c) $\underset{\underset{Br}{|}}{CH_2}-\underset{\underset{Cl}{|}}{CH_2}$

b) $CH_3-\overset{\overset{Cl}{|}}{CH}-\overset{\overset{Cl}{|}}{CH_2}$

d) $CH_3-CH_2\underset{}{\overset{\overset{CH_3}{|}}{CH}} - \overset{\overset{Cl}{|}}{CH}-CH_3$

12. Write structural formulas for each of the following halogenated hydrocarbons:

a) 1-bromo-3-methylpentane
b) 2,2-dichloropropane
c) 2,2-dibromo-3-chloro-3-methylheptane
d) 1,1,1-trichlorobutane

13. Name the following alcohols using IUPAC rules:

a) $CH_3-\underset{\underset{OH}{|}}{CH}-\overset{\overset{CH_3}{|}}{CH}-CH_3$

c) $CH_3-CH_2-\underset{\underset{CH_3}{\overset{|}{\underset{|}{CH_2}}}}{\overset{\overset{CH_3}{|}}{C}} - \overset{\overset{OH}{|}}{CH}-CH_2-CH_3$

b) $CH_3-\overset{\overset{CH_3}{|}}{CH}-CH_2-\overset{\overset{OH}{|}}{CH_2}$

d) $CH_3-\underset{\underset{OH}{|}}{\overset{\overset{CH_3}{|}}{C}}-CH_3$

14. Write structural formulas for the following alcohols:

a) 2,3-dimethyl-3-pentanol   c) 2-methylcyclohexanol
b) 2,3-butanediol   d) 1-butanol

15. Assign a common and an IUPAC name to each of the following ethers:

a) $CH_3-O-CH_2-CH_2-CH_3$ c) CH_3-O-CH_3

b) $CH_3-CH_2-O-CH_2-CH_3$ d) $CH_3-CH_2-CH_2-CH_2-O-CH_2-CH_3$

16. Write structural formulas for the following ethers:

a) ethoxyethane   c) 2-methoxybutane
b) methoxypropane   d) diisopropyl ether

17. Using IUPAC rules, name the following carboxylic acids:

a) $CH_3-CH_2-CH-C\overset{O}{\underset{OH}{\diagup}}$
   $\quad\quad\quad\quad | $
   $\quad\quad\quad CH_3$

c) $\quad\quad CH_3$
   $\quad\quad\quad | \quad\quad O$
   $CH_3-C-C\overset{}{\underset{OH}{\diagup}}$
   $\quad\quad\quad |$
   $\quad\quad\quad CH_3$

b) $CH_3CH-CH_2-C\overset{O}{\underset{OH}{\diagup}}$
   $\quad\quad | $
   $\quad\quad CH_3$

d) $\quad\quad\quad\quad CH_3 \quad CH_3$
   $\quad\quad\quad\quad\quad | \quad\quad\quad | \quad\quad O$
   $CH_3-CH_2-CH - CH-C\overset{}{\underset{OH}{\diagup}}$

18. Draw structural formulas for each of the following carboxylic acids:

a) hexanoic acid
b) 4-ethylhexanoic acid
c) propanoic acid
d) methanoic acid

19. Using IUPAC rules, name the following esters:

a) $CH_3-CH-CH_2-\overset{O}{\overset{||}{C}}-O-CH_2-CH_3$
   $\quad\quad\quad | $
   $\quad\quad\quad CH_3$

c) $H-\overset{O}{\overset{||}{C}}-O-CH_3$

b) $CH_3-\overset{O}{\overset{||}{C}}-O-CH_2-CH_3$

d) $CH_3-CH_2-O-\overset{O}{\overset{||}{C}}-H$

20. Draw structural formulas for each of the following esters:

a) ethyl butanoate
b) methyl propanoate
c) propyl ethanoate
d) isopropyl ethanoate

MULTIPLE CHOICE EXERCISES

1. Which of the following statements concerning saturated hydrocarbon molecules is <u>incorrect</u>:

   a) only two elements are present
   b) all bonds present are single bonds
   c) all carbon atoms must have 4 bonds
   d) all carbon atoms must be bonded to 3 hydrogen atoms

2. All noncyclic alkanes have molecular formulas which fit the general formula:

   a) C_nH_{2n}
   b) $C_{2n}H_n$
   c) C_nH_{2n+2}
   d) $C_{2n+2}H_n$

3. Which of the following statements concerning hydrocarbon isomers is correct?

   a) they must contain the same number of carbon atoms but may differ in the number of hydrogen atoms present
   b) they must differ in the number of hydrogen atoms present
   c) they must contain the same number of carbon atoms and also the same number of hydrogen atoms
   d) they must contain differing numbers of carbon atoms

4. Structural isomerism is possible in alkanes only if:

   a) 3 or more carbon atoms are present
   b) 4 or more carbon atoms are present
   c) 6 or more carbon atoms are present
   d) 8 or more carbon atoms are present

5. What is the IUPAC name for

$$
\begin{array}{c}
\text{C} \\
| \\
\text{C-C-C-C-C-C} \\
| \\
\text{C}
\end{array}
$$

   a) 3-methylhexane
   b) 3-dimethylhexane
   c) 3-methyl-3-methylhexane
   d) 3,3-dimethylhexane

6. When the molecular formulas for cyclic and noncyclic alkanes with the same number of carbons are compared, it is always found that the cycloalkane has:

   a) 2 more hydrogen atoms
   b) the same number of hydrogens as the noncyclic alkane
   c) 2 less hydrogen atoms
   d) 4 less hydrogen atoms

7. When a geometrical figure (hexagon) is used to represent cyclohexane, it is assumed that each vertice of the figure represents a carbon atom and:

   a) one hydrogen atom
   b) two hydrogen atoms
   c) three hydrogen atoms
   d) four hydrogen atoms

8. A cycloalkene having only one double bond will fit the general formula:

   a) C_nH_{2n-4}
   b) C_nH_{2n-2}
   c) C_nH_{2n}
   d) C_nH_{2n+2}

9. The bonding in benzene differs from that in aliphatic hydrocarbons in that:

   a) some of the electrons are involved in delocalized bonding
   b) carbon atoms may have 5 bonds
   c) carbon atoms may form only 3 bonds
   d) the electrons have aromas associated with them

10. Two adjacent substituents on a benzene ring are said to be:

    a) ortho to each other
    b) meta to each other
    c) para to each other
    d) nexo to each other

11. Which of the following is not a possible dimethylbenzene isomer:

    a)

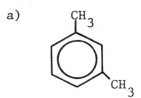

    b)

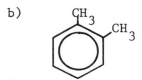

c)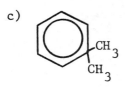

d)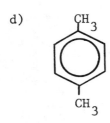

12. Crude petroleum is:

    a) a hydrocarbon mixture obtained from underground
       "liquid pools"
    b) a mixture of methane and ethane
    c) a complex mixture of many hydrocarbons
    f) over 80% methane

13. The IUPAC name for the compound

$$\begin{array}{cc} Br & C \\ | & | \\ C-C - & C-C \end{array}$$ is

    a) 2-bromopentane
    b) 2-bromobutane
    c) 2-bromo-3-methylbutane
    d) bromomethylbutane

14. The IUPAC name for the compound below is:

$$CH_3-CH_2-\underset{\underset{OH}{|}}{CH}-CH_3$$

    a) hydroxybutane
    b) 2-butanol
    c) 1,1-hydroxymethylpropane
    d) n-butyl alcohol

15. In which of the following pairs of alcohols does each
    alcohol contain the same number of -OH groups?

    a) rubbing alcohol, glycerin
    b) wood alcohol, grain alcohol
    c) ethylene glycol, drinking alcohol
    d) glycerin, ethylene glycol

16. Which of the following is an ether?

a) $CH_3-O-\overset{\overset{\displaystyle O}{\displaystyle \|}}{C}-CH_3$

b) CH_3-O-CH_3

c) CH_3-CH_2-Br

d) $CH_3-CH_2-\overset{\overset{\displaystyle O}{\displaystyle \|}}{C}-OH$

17. The functional group present in acids contains:

a) a carbon–hydrogen bond
b) both a carbon–oxygen single bond and a carbon–oxygen double bond
c) only the elements carbon and oxygen
d) only the elements oxygen and hydrogen

18. Which of the following can be considered to be organic derivatives of water in which both hydrogens have been replaced by hydrocarbon groups:

a) esters
b) ethers
c) carboxylic acids
d) alcohols

19. The pleasant, characteristic odor of fruit flavorings is often associated with the presence of:

a) ethers
b) esters
c) carboxylic acids
d) alcohols

20. Which of the following is an ester?

a) 2-butanol
b) methoxymethane
c) ethylbutanoate
d) 2-bromoheptane

ANSWERS TO PROBLEMS SET

1. C–C–C–C–C   pentane

 C–C–C–C   2-methylbutane
 |
 C

 C
 |
 C–C–C   2,2-dimethylpropane
 |
 C

2. a) 3-ethylpentane   c) 2,2,3,6-tetramethyloctane
 b) 2,2-dimethylpropane   d) 3,5-dimethyloctane

3. a) CH_3–CH–CH_2–CH_2–CH_2–CH_3 c)
 |
 CH_3

$$CH_3-CH_2-\overset{\overset{\displaystyle CH_3}{|}}{\underset{\underset{\displaystyle CH_3}{|}}{C}}-\overset{\overset{\displaystyle CH_2}{|}}{\underset{}{CH}}-CH_2-CH_3$$

 b) CH_3–CH – CH–CH_3 d)
          | |
         CH_3 CH_3

$$CH_3-\overset{\overset{\displaystyle CH_3}{|}}{\underset{\underset{\displaystyle CH_3}{|}}{C}}-CH_2-\overset{\overset{\displaystyle CH_3}{|}}{CH}-CH_3$$

4. a) ethylcyclopropane
 b) 1,2-dimethylcyclobutane
 c) 1-ethyl-3-methylcyclopentane
 d) 1,3-dimethylcyclohexane

5. a) 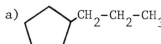 CH_2–CH_2–CH_3 c)

 b) CH_3
           |
         CH–CH_3 d) CH_3 CH_3

6. a) 2-methyl-2-butene   d) 2-pentyne
 b) 3-methyl-1-butene   e) 2,3-dimethyl-2-butene
 c) 1,3-butadiene   f) 4-methyl-1-cyclopentene

7. a) $CH_2=CH-CH_2-CH_3$

    d)
    $$CH_3-C{\equiv}C-\underset{\underset{CH_3}{|}}{\overset{\overset{CH_3}{|}}{C}}-\underset{\overset{CH_3}{|}}{CH}-CH_3$$

    b) $CH_3-CH=CH-\underset{\underset{\underset{CH_3}{|}}{\overset{|}{CH_2}}}{CH}-CH_2-CH_2-CH_3$

    e) $CH_2=CH-CH=CH-CH_3$

    c) $CH{\equiv}C-\underset{\overset{|}{CH_3}}{CH}-CH_2-CH_3$

    f)

8. a) 1,2-dimethylbenzene or o-dimethylbenzene or o-xylene
    b) 1,2,3-trimethylbenzene
    c) 1-ethyl-3-methylbenzene or m-ethylmethylbenzene
    d) isopropylbenzene
    e) 1-methyl-2-propylbenzene or o-methylpropylbenzene
    f) 1,4-diethylbenzene or p-diethylbenzene

9. a)

    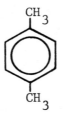

    d)

    b)

    e)

    c)

    f)

10. a) halogenated hydrocarbon   d) carboxylic acid
    b) alcohol   e) ester
    c) ether

11. a) 2-bromobutane   c) 1-bromo-2-chloroethane
    b) 1,2-dichloropropane   d) 2-chloro-3-methylpentane

12. a) $CH_2-CH_2-CH-CH_2-CH_3$
        Br   CH$_3$

    c) $CH_3-\overset{Br}{\underset{Br}{C}} - \overset{CH_3}{\underset{Cl}{C}}-CH_2-CH_2-CH_2-CH_3$

    b) $CH_3-\overset{Cl}{\underset{Cl}{C}}-CH_3$

    d) $Cl-\overset{Cl}{\underset{Cl}{C}}-CH_2-CH_2-CH_3$

13. a) 3-methyl-2-butanol
    b) 3-methyl-1-butanol
    c) 3-ethyl-3-methyl-4-hexanol
    d) 2-methyl-2-propanol

14. a) $CH_3-\overset{CH_3}{CH} - \overset{OH}{\underset{CH_3}{C}}-CH_2-CH_3$

    c) (cyclohexane ring with OH and CH$_3$)

    b) $CH_3-\underset{OH}{CH}-\underset{OH}{CH}-CH_3$

    d) $HO-CH_2-CH_2-CH_2-CH_3$

15. a) methoxypropane, methyl propyl ether
    b) ethoxyethane, diethyl ether
    c) methoxymethane, dimethyl ether
    d) ethoxybutane, ethyl n-butyl ether

16. a) $CH_3-CH_2-O-CH_2-CH_3$

    c) $CH_3-\underset{\underset{CH_3}{O}}{CH}-CH_2-CH_3$

    b) $CH_3-O-CH_2-CH_2-CH_3$

    d) $CH_3-\underset{CH_3}{CH}-O-\underset{CH_3}{CH}-CH_3$

17. a) 2-methylbutanoic acid   c) 2,2-dimethylpropanoic acid
    b) 3-methylbutanoic acid   d) 2,3-dimethylpentanoic acid

18. a) $CH_3-CH_2-CH_2-CH_2-CH_2-C\overset{O}{\underset{OH}{\diagup}}$ c) $CH_3-CH_2-C\overset{O}{\underset{OH}{\diagup}}$

b) $CH_3-CH_2-\underset{\underset{\underset{CH_3}{|}}{\overset{|}{CH_2}}}{CH}-CH_2-CH_2-C\overset{O}{\underset{OH}{\diagup}}$ d) $H-C\overset{O}{\underset{OH}{\diagup}}$

19. a) ethyl-3-methylbutanoate   c) methylmethanoate
     b) ethylethanoate   d) ethylmethanoate

20. a) $CH_3-CH_2-O-\overset{O}{\overset{||}{C}}-CH_2-CH_2-CH_3$ c) $CH_3-CH_2-CH_2-O-\overset{O}{\overset{||}{C}}-CH_3$

b) $CH_3-O-\overset{O}{\overset{||}{C}}-CH_2-CH_3$ d) $CH_3-\underset{\underset{CH_3}{|}}{CH}-O-\overset{O}{\overset{||}{C}}-CH_3$

ANSWERS TO MULTIPLE CHOICE EXERCISES

1.	d	11.	c
2.	c	12.	c
3.	c	13.	c
4.	b	14.	b
5.	d	15.	b
6.	c	16.	b
7.	b	17.	b
8.	b	18.	b
9.	a	19.	b
10.	a	20.	c